John Elliot

An Account of the Nature and medicinal Virtues of the principal

Mineral Waters

Of Great Britain and Ireland

John Elliot

An Account of the Nature and medicinal Virtues of the principal Mineral Waters
Of Great Britain and Ireland

ISBN/EAN: 9783337143954

Printed in Europe, USA, Canada, Australia, Japan

Cover: Foto ©ninafisch / pixelio.de

More available books at **www.hansebooks.com**

A N

A C C O U N T

O F   T H E

Nature and Medicinal Virtues

O F   T H E

Principal Mineral Waters

O F

GREAT BRITAIN and IRELAND;

A N D   T H O S E

MOST IN REPUTE ON THE CONTINENT.

T O   W H I C H   A R E   P R E F I X E D

Directions for Impregnating Water with Fixed Air, in order to communicate to it the peculiar Virtues of Pyrmont Water, and other Mineral Waters of a similar Nature, extracted from Dr. Priestley's Experiments on Air.

WITH AN APPENDIX,

Containing a Description of Dr. Nooth's Apparatus, with the Improvements made in it by others. And a Method of Impregnating Water with Hepatic Air, so as to imitate the Aix-la-Chapelle and other Sulphureous Waters.

By J O H N   E L L I O T, M. D.

The Second Edition, corrected and enlarged.

L O N D O N:

PRINTED FOR J. JOHNSON, Nᵒ. 72, ST. PAUL'S CHURCH-YARD, MDCCLXXXIX.

ADVERTISEMENT

TO THE

FIRST EDITION.

DR. PRIESTLEY's Pamphlet on the Impregnation of Water with Fixed Air being out
of print, and that Gentleman having no intention
of republishing it, I have judged proper to prefix
it to the following Tract, with the additions, as
printed in his second Volume of Experiments on
Air. This was done as well that the reader might
be entertained with the history of the discovery, as
instructed in an easy method of making the impregnation when Dr. Nooth's apparatus might
not be at hand.

J. E.

Newman Street,
Aug. 30, 1781.

———————————

IN this Second Edition the contents of the
principal Waters, and their Proportions, when
they could be obtained from any good authority,
have been inserted: some, which have come into
repute since the publication of the former edition,
are added: the proportions of the ingredients for
imitating different Mineral Waters have been
altered, to bring them nearer to what, from the
analyses of later chemists, we may presume to be
their true composition: the method of preparing
the mephitic alkaline water is given in a more
full and improved manner; and that part which
relates to the analysis of waters is considerably
enlarged.

C O N T E N T S

OF THE IMPREGNATION OF WATER WITH
FIXED AIR.

AN ACCOUNT OF THE NATURE, PROPERTIES, AND MEDICINAL VIRTUES OF THE PRINCIPAL MINERAL WATERS IN GREAT BRITAIN AND IRELAND.

IMPREGNATION OF WATER

.WITH

FIXED AIR.

From Dr. PRIESTLEY's Experiments, Vol. II,

CHAPTER I.

THE HISTORY OF THE DISCOVERY.

IT often amufes me when I review the hiftory of experimental philo-fophy, to obferve how very nearly one difcovery is connected with another, and yet that, for a long time, no per-fon fhall have perceived that con-nection, fo as to have been actually led from the one to the other ; and efpecially that he who made the firft difcovery fhould ftop fhort in his pro-grefs, and not advance a fingle ftep

farther,

farther, to make the other, which
was perhaps of infinitely more confequence. And yet the cafe may be
fuch, that it fhall be fo far from re-
quiring more genius, or ingenuity,
to advance that other ftep, that it is
rather a matter of wonder, how it
was poffible for the moft common ca-
pacity to ftop fhort of it. We alfo
frequently find that they who make
the moft important philofophical difcoveries overlook the moft obvious
ufes of them. Several ftriking ex-
amples of this kind will be found in
my *Hiftory of electricity*, and alfo in
the *Hiftory of difcoveries relating to
vifion, light, and colours.*

In fuch cafes as thefe it behoves an
hiftorian to be much on his guard,
left he fhould haftily conclude that to
have been fact which he only *imagines*
muft have been fo, but for which no di-
rect evidence can be produced. As this

†   is

is a cafe of fome curiofity refpecting the human mind, I fhall give an inftance of it; and I am able to produce a very remarkable one relating to the fubject of this fection.

When it was difcovered that the acidulous tafte and peculiar virtues of Pyrmont water, and other mineral waters of a fimilar nature, were owing to the fixed air which they contained; when this air had been actually expelled from the water, and it was found that the fame water, and even other water, would reimbibe the fame air; we are apt to conclude that the perfon who made thefe difcoveries, and efpecially the laft of them (who alfo muft have known that fixed air is a thing very eafy to be procured) muft have immediately gone to work to reduce this *theory* into *practice*, by actually impregnating common water with fixed air, in order to give it the

peculiar

peculiar virtues of thofe medicinal mineral waters which are fo highly, and fo juftly valued, and which are procured at fo great an expence, efpecially in this country. Accordingly, Dr. Nooth has advanced, Phil. Tranf. vol. lxv. p. 59, that " the " poffibility of impregnating water " with fixed air was no fooner afcer- " tained by experiment, than various " methods were contrived to effect the " impregnation;" and I doubt not this ingenious philofopher impofed upon himfelf in the manner defcribed above. This, however, is fo far from being the cafe, that I do not believe it is poffible to produce the leaft evidence that any perfon had the thing in view before the publication of my pamphlet upon that fubject, in the year 1772.

Indeed had this thing been fo much as *an object of attention* to philofo-

phers,

phers, it is impoffible but that fome of them muft have hit upon a method that would have fufficiently fucceeded. Nay, the thing is fo very eafy, and the end attainable in fo many ways, that there muft have been, in a very fhort time, a great variety of methods to impregnate water with fixed air, as there are now; and we fhould certainly have heard of *artificial mineral waters* being made according to them. It is impoffible not to conclude fo, when we confider the *time that has elapfed* fince the publication of all the difcoveries that led to it.

Dr. Brownrigg's paper, giving an account of his difcovery of fixed air in the Spa water, was read at the Royal Society June the 13th 1765, and was publifhed in 1766. This excellent philofopher compleatly decompofed that mineral water, but he gives no hint of his having fo much

as

as attempted to *recompofe* it, or of making a fimilar water, by impregnating common water with the fame volatile principle. It is fufficiently evident that he had not thought of this, though we may wonder that he fhould not have done it, becaufe he has not mentioned it, as an object of purfuit.

In the year following, Mr. Cavendifh's valuable papers on the fubject of factitious air were publifhed. He firft afcertained how much fixed air a given quantity of water could be made to imbibe ; yet it does not appear that he ever thought of *tafting* the water, much lefs that he thought of making any *practical ufe* of his difcovery.

If any negative argument can be decifive, it is that in the year 1772, the very year in which my pamphlet came out, Dr. Falconer publifhed

his

his excellent and elaborate treatife on the *Bath waters,* in which he treats very largely of mineral waters in general, and all their poffible impregnations ; and yet, though he treats of *fixed air* as one ingredient in many of them, fee p. 185, he drops no hint about compofing fuch water, by imparting fixed air to common water. Alfo on the 12th of September in the fame year, Dr. Rutherford publifhed his ingenious *Differtation on Fixed Air,* in which he fpeaks of the prefence of it in Pyrmont water, p. 3, but without giving the leaft hint of his being acquainted with any method of imitating them. And yet, Dr. Nooth fays, in fact, that from the year 1766, at the lateft, *various methods* were contrived to effect the impregnation, though he allows that I was the only perfon who " publifhed

B 4 " any

" any defcription of an apparatus cal-
" culated entirely for this purpofe."

According to this account of the
matter there were, in the interval between 1766 and 1772, a fpace of fix
years, a variety of methods for impregnating water with fixed air, fome
of them prior to, and perhaps much
better than mine (though he gives no
hint of his own having been invented
in that period, but fpeaks of it as fug-
gefted by the confideration of the im-
perfection of mine) but that I happened to get the ftart in the publication. Dr. Falconer, however, though
the friend of Dr. Nooth (fee his treatife on Bath Water, vol. ii. p. 323)
had certainly never heard of any of
thofe methods, or even of mine, at
the very termination of that period;
and though my own acquaintance
with philofophical and medical people

ple is pretty extenſive, I never heard of any of the *various methods* that Dr. Nooth ſpeaks of; nor ſince the publication of my method have I heard of any perſon whatever having pretended to have done the ſame thing before; though nothing is more common than ſuch claims, and very often on the moſt trifling pretences.

Mr. Venelle, indeed, immediately upon the tranſlation of my pamphlet into French, which was within a few weeks after the publication of it in Engliſh (owing to the laudable zeal of Mr. Trudaine, for promoting all philoſophical and uſeful improvements) publiſhed an extract of his papers from the *Memoires de Mathematique & de Phyſique*, to vindicate to himſelf not my diſcovery, but, in fact, that of Dr. Brownrigg. However, what he pretends to have diſcovered was, that the virtues of the acidulous wa-

B 5 ters

ters were owing to *air, in general,* without having any idea of the difference between fixed air and common air; fo that his difcovery was fo far from being the fame with mine, that it could not poffibly have led into it.

As I have hitherto only publifhed the method of impregnating water with fixed air in a fmall pamphlet, for the ufe of thofe who might chufe to reduce it into practice, without giving any account of the manner in which the difcovery (if it deferves to be called one) was made, which has been my cuftom with refpect to every thing elfe, I fhall do it here; and I hope the narrative will not be altogether difpleafing, as this bufinefs has gained fo much attention in all parts of Europe, as well as in England, and promifes in a fhort time to fave the very great expence of tranfporting

acidulous

acidulous waters to confiderable dif-
tances, by fuperceding, in a great
meafure, the ufe of them. And
though what I have done in this bufinefs has certainly the leaft merit poffible with refpect to *ingenuity*, I fhall
always confider it as one of the *hap-
piest* thoughts that ever occurred to
me; becaufe it has proved to be of
very fignal *benefit* to mankind, and
will, I doubt not, be of much more
confequence in a courfe of time.

It was a little after Midfummer in
1767, that I removed from Warring-
ton to Leeds; and living, for the firft
year, in a houfe that was contiguous
to a large common brewery, fo good
an opportunity produced in me an inclination to make fome experiments
on the fixed air that was conftantly
produced in it. Had it not been for
this circumftance, I fhould, probably,
never have attended to the fubject of

air

air at all. Happening to have read Dr. Brownrigg's excellent paper on the Spa water about the fame time, one of the firft things that I did in this brewery was to place fhallow veffels of water within the region of fixed air, on the furface of the ferment-ing veffels ; and having left them all night, I generally found, the next morning, that the water had acquired a very fenfible and pleafant impreg-nation ; and it was with peculiar fa-tisfaction that I firft drank of this water, which I believe was the firft of its kind that had ever been tafted by man.

This procefs, however, was very flow. But after fome time it occur-red to me, that the impregnation might be accelerated, by pouring the water from one veffel into another, while they were both held within the fphere of the fixed air ; and accord-ingly

ingly I found that I could do as much in about five minutes in this way, as I had been able to do in many hours before. Several of my friends who vifited me while I lived in that houfe will remember my taking them into that brewery, and giving them a glafs of this artificial Pyrmont water, made in their prefence. Among others, I will take the liberty to mention John Lee, Efq; of Lincoln's Inn, who was particularly ftruck with the contrivance, and the effect of it. This was in the fummer of the year 1768.

One would naturally think, that having actually impregnated common water with fixed air, produced in a brewery, I fhould immediately have fet about doing the fame thing with air fet loofe from chalk, &c. by fome of the ftronger acids; and I do remember that it did occur to me that the thing was poffible. But, eafy as

the

the practice proved to be, no method of doing it at that time occurred to me. I still continued to make my Pyrmont water in the manner above mentioned 'till I left that situation, which was about the end of the summer 1768 ; and from that time, being engaged in other similar pursuits, with the result of which the public are acquainted, I made no more of the Pyrmont water 'till the spring of the year 1772.

In the mean time I had acquainted all my friends with what I had done, and frequently expressed my wishes that persons who had the care of large *distilleries* (where I was told that fermentation was much stronger than in common breweries) would contrive to have vessels of water suspended within the fixed air, which they produced, with a farther contrivance for agitating the surface of the water ; as I did

I did not doubt but that, by this means, they might, with little or no expence, make great quantities of Pyrmont water; by which they might at the fame time both ferve the Public, and benefit themfelves. For I never had the moft diftant thought of making any advantage of the fcheme myfelf.

In all this time, viz. from 1767 to 1772, I never heard of any method of impregnating water with fixed air but that above mentioned. My thinking at all of reducing to practice any method of effecting this, by air diflodged from chalk, and other calcareous fubftances, was owing to a mere accident. Being at dinner with the Duke of Northumberland, in the fpring of the year laft mentioned, his Grace pro-.duced a bottle of water diftilled by Dr. Irving for the ufe of the navy. This water was perfectly fweet, but,

like

like all diftilled water, wanted the
brifknefs and fpirit of frefh fpring water; when it immediately occurred to
me that I could eafily mend that water for the ufe of the navy, and perhaps fupply them with an eafy and
cheap method of preventing or curing
the fea-fcurvy, viz. by impregnating
it with fixed air.   For having been
bufy about a year before with my experiments on air, in the courfe of
which I had afcertained the propor-
tional quantity of feveral kinds of air
that given quantities of water would
take up, I was at no lofs for the *me-*
thod of doing it in general, viz. in-
verting a jar filled with water, and
conveying air into it from bladders
previoufly filled with air.   This
fcheme I immediately mentioned to
the Duke and the company, who all
feemed to be much pleafed with it,
and expreffed their wifhes that I would
attend

attend to it, and endeavour to reduce it into practice; which I promifed to do.

The next day I provided a fmall apparatus, adapted to this purpofe, at my lodgings, which was very eafy, as it required no other veffels but fuch as are in conftant family ufe, and with this I prefently impregnated a quantity of the New River water, fo as to make it imbibe about its bulk of air. But I was far from having hit upon the *eafieft method* of doing it; for my jars were of an equal width throughout. However, with thefe veffels the procefs was compleated in about twenty minutes, or half an hour.

A few days after this, having an invitation to wait upon Sir George Savile, I carried with me a bottle of my impregnated water, and told him the ufe that might be made of it, viz. that of fupplying a pleafant and wholefome

fome beverage for feamen, and fuch as might probably prevent or cure the fea-fcurvy. Sir George, with that warmth with which he efpoufes every thing that he conceives to be for the public good, infifted upon writing a card immediately to Lord Sandwich, propofing to introduce me to him, as having *a propofal for the ufe of the navy.* As I could make no objection, the card was accordingly written, and an anfwer was prefently returned from his Lordfhip, informing us that he would be glad to fee us the next day. Upon this I drew up fomething in the form of a *propofal*, which, accompanied by Sir George, I prefented to his Lordfhip, who promifed to lay it before the Board of Admiralty.

Prefently after this I had notice from the Secretary to the Board of Admiralty, that the *College of Phyficians* were appointed to examine my
—   propofal,

propofal, and to make their report of
it to the Board, and an early day was
fixed for me to wait upon them at
their hall in Warwick-Lane; where,
before a very full meeting, I produced
a bottle of my impregnated water,
and alfo, at their requeft, fetched my
apparatus, and fhewed them the manner in which I had impregnated it.
There were prefent feveral of the moft
eminent phyficians in London; but
both the *fcheme* and the *object* of it,
appeared to be entirely new to every
one of them; and moft of them feemed to be much pleafed with it.

Accordingly, a favourable report
was made to the Board of Admiralty,
and I was acquainted by the Secretary, that the Captains of the two
fhips which were juft then failing for
the South-Seas had orders to make a
trial of the impregnated water; and
for their ufe I drew out my *Directions*

in

in writing, and fent a drawing of the neceffary apparatus. The method which I had now got into was a great improvement upon that which I had made ufe of before the College of Phyficians. For, in confequence of giving more attention to it, I had, by that time, brought it to the ftate in which it is defcribed in the pamphlet.

In the mean time, I had, before I left London, in the fpring of that year, made the experiment of the impregnation of water with fixed air in the prefence of moft of my philofophical acquaintance, and their friends, both at my own lodgings, and in other places. But upon none of thefe occafions did it appear that any of them had heard of any other perfon having had the fame thing in view.

Laftly, I will obferve, that Sir John Pringle, in his *Difcourfe on different kinds*

kinds of air (in which he has, with the greateſt exactneſs, aſſigned to every perſon concerned in theſe diſcoveries their due ſhare of praiſe) gives no hint of his being acquainted with any other method of impregnating water with fixed air, than that which I had publiſhed. He certainly had not heard of any of thoſe to which Dr. Nooth alludes.

As I have not to this day, directly or indirectly, made the leaſt advantage of this ſcheme; but, on the contrary, am juſt ſo much a loſer by it as the experiments coſt me, I think it is not too much for the Public to allow me, what I believe is ſtrictly my due, *the ſole merit of the diſcovery*; which with reſpect to *ingenuity*, or ſagacity, is next to nothing; but with reſpect to its *utility* is, unqueſtionably, of un-ſpeakable value to my country and to mankind.

CHAP-

CHAPTER II.

DIRECTIONS FOR IMPREGNATING
WATER WITH FIXED AIR.

Sect. 1. *The Preface to the Directions
as first published.*

THE method of impregnating
water with fixed air, of which a defcription is given in this pamphlet, I
hit upon in a courfe of experiments;
an account of which was lately communicated to the Royal Society; containing obfervations on feveral different kinds of air, with only a hint of
the method of combining this particular kind with water or other fluids.
Judging that water thus impregnated
with fixed air muft be particularly ferviceable in long voyages, by preventing or curing the fea-fcurvy, according to the theory of Dr. Macbride;

and

and all the Phyficians of my acquaintance concurring with me in that opinion, I made the firft communication
of it to the Lords of the Admiralty,
who referred me fo the College of
Phyficians; and thofe gentlemen being pleafed to make a report favourable to the fcheme, a trial has been
ordered to be made of it on board
fome of his Majefty's fhips. To make
this procefs more generally known,
and that more frequent trials may be
made by water thus medicated, at land
as well as at fea, I have been induced
to make the prefent publication.

Sir John Pringle firft obferved, that
putrefaction was checked by fermentation ; and Dr. Macbride difcovered
that this effect was produced by the
fixed air which is generated in that
procefs, and upon that principle recommended the ufe of *wort*, as fupplying a quantity of this fixed air, by
fermen-

fermentation in the ftomach, in the fame manner as it is done by frefh vegetables, for which he, therefore, thought that it would be a fubftitute; and experience has confirmed his conjecture.   Dr. Black found that lime - ftone, and all calcareous fub-ftances, contain fixed air, that the pre-fence of it makes them what is called *mild*, and that the deprivation of it renders them *cauftic*; Dr. Brownrigg farther difcovered that Pyrmont, and other mineral waters, which have the fame acidulous tafte, contain a confi-derable proportion of this very kind of air, and that upon this their pecu-liar fpirit and virtues depend; and I think myfelf fortunate in having hit upon a very eafy method of commu-nicating this air to any kind of water, or, indeed, to almoft any fluid fub-ftance.   In fhort, by this method this great antifeptic principle may be ad-

miniftered

miniftered in a great variety of agreeable vehicles.

If this difcovery (though it doth
not deferve that name) be of any ufe
to my countrymen, and to mankind
at large, I fhall have my reward. For
this purpofe I have made the communication as early as I conveniently
could, fince the lateft improvements
that I have made in the procefs; and
I cannot help expreffing my wifhes,
that all perfons, who difcover any
thing that promifes to be generally
ufeful, would adopt the fame method.

Sect. 2.   *The Directions.*

If water be only in contact with
fixed air, it will begin to imbibe it,
but the mixture is greatly accelerated
by agitation, which is continually
bringing frefh particles of air and water into contact.   All that is neceffary, therefore, to make this procefs

C   expeditious

expeditious and effectual, is firſt to
procure a ſufficient quantity of this
fixed air, and then to contrive a method by which the air and water may
be ſtrongly agitated in the ſame veſſel,
without any danger of admitting the
common air to them ; and this is eaſily done by firſt filling any veſſel with
water, and introducing the fixed air
to it, while it ſtands inverted in another veſſel of water. That every
part of the proceſs may be as intelli-
gible as poſſible, even to thoſe who
have no previous knowledge of the
ſubject, I ſhall deſcribe it very mi-
nutely, ſubjoining ſeveral remarks and
obſervations relating to varieties in the
proceſs, and other things of a miſ-
cellaneous nature.

The Preparation.

Take a glaſs veſſel, *a*, fig. 1.
with a pretty narrow neck, but ſo
formed,

formed, that it will ftand upright with its mouth downwards, and having filled it with water, lay a flip of clean paper, or thin pafteboard, upon it. Then, if they be preffed clofe together, the veffel may be turned upfide down, without danger of admitting common air into it; and when it is thus inverted, it muft be placed in another veffel, in the form of a bowl or bafon, *b*, with a little water in it, fo much as to permit the flip of paper or pafteboard to be withdrawn, and the end of the pipe *c* to be introduced.

This pipe muft be flexible, and airtight, for which purpofe it is, I believe, beft made of leather, fewed with a waxed thread, in the manner ufed by fhoe-makers. Into both ends of this pipe a piece of a quill fhould be thruft, to keep them open, while one of them is introduced into the

C 2　　　　veffel

veſſel of water, and the other into the
bladder *d*, the oppoſite end of which
is tied round a cork, which muſt be
perforated, the hole being kept open
by a quill; and the cork muſt fit a
phial *e*, two thirds of which ſhould
be filled with chalk juſt covered with
water.

I have ſince, however, found it moſt
convenient to uſe a *glaſs tube*, and to
preſerve the advantage which I had,
of agitating the veſſel *e*, I have *two
bladders*, communicating by a perforated cork, to which they are both
tied.   For one bladder would hardly
give room enough for that purpoſe.

The Proceſs.

Things being thus prepared, and
the phial containing the chalk and
water being detached from the bladder,
and the pipe alſo from the veſſel of
water, pour a little oil of vitriol upon
the

the chalk and water; and having carefully preffed all the common air out of the bladder, put the cork into the bottle prefently after the effervefcence has begun. Alfo prefs the bladder once more after a little of the newly generated air has got into it, in order the more effectually to clear it of all the remains of the common air; and then introduce the end of the pipe into the mouth of the veffel of water as in the drawing, and begin to agitate the chalk and water brifkly. This will prefently produce a confiderable quantity of fixed air, which will diftend the bladder; and this being preffed, the air will force its way through the pipe, and afcend into the veffel of water, the water at the fame time defcending, and coming into the bafon.

When about one half of the water is forced out, let the operator lay his

C 3   hand

hand upon the uppermoft part of the veffel, and fhake it as brifkly as he can, not to throw the water out of the bafon ; and in a few minutes the water will abforb the air ; and taking its place, will nearly fill the veffel as at the firft.   Then fhake the phial containing the chalk and water again, and force more air into the veffel, 'till, upon the whole, about an equal bulk of air has been thrown into it.   Alfo fhake the water as before, 'till no more of the air can be imbibed.   As foon as this is perceived to be the cafe, the water is ready for ufe ; and if it be not ufed immediately, fhould be put into a bottle as foon as poffible, well corked, and cemented.   It will keep, however, very well, if the bottle be only well corked, and kept with the mouth downwards.

Obferva-

Obfervations.

1. The bafon may be placed inverted upon the veffel full of water, with a flip of paper between them, and then both turned upfide down together; but all this trouble will be faved by having a larger veffel of water, in which both of them may be immerfed.

2. If the veffel containing the water to be agitated be large, it may be moft convenient fit... to place it inverted, in a bafon full of water, and then to draw out the common air by means of a fyphon, either making ufe of a fyringe, or drawing it out with the mouth. In this cafe, alfo, fome kind of handle fhould be faftened to the bottom of the veffel, for the more eafy agitation of it.

3. A narrow-mouthed veffel is not neceffary, but it is the moft proper for

C 4 the

the purpofe, becaufe it may be agitated with lefs danger of the common air getting into it.

4. The flexible pipe is not necef-fary, though I think it is exceedingly convenient. When it is not ufed, a bent tube, *a*, fig. 2. (for which glafs is the moft proper) muft be ready to be inferted into the hole made in the cork, when the bladder containing the fixed air is feparated from the phial, in which it was generated. The extremity of this tube being put un-der the veffel of water, and the blad-der being compreffed, the air will be conveyed into it, as before.

5. If the ufe of a bladder be ob-jected to, though nothing can be more inoffenfive, the phial containing the chalk and water muft not be agitated at all, or with the greateft caution; unlefs a fmall phial, *a*, fig. 3. be in-terpofed between the phial and the
veffel

veffel of water, in the manner repre-
fented in the drawing. For by this
means the chalk and water that may
be thrown up the tube *b* will lodge at
the bottom of the phial *a*, while nothing but the air will get into the
pipe *c*, and fo enter the water. If the
tube *b* be made of tin or copper, the
fmall phial *a* will not need any other
fupport, the cork into which the ex-
tremities of both the tubes are inferted being made to fit the phial very
exactly.

6. The phial *e*, fig. 1. fhould always be placed, or held, confiderably
lower than the veffel *a* ; that if any
part of the mixture fhould be thrown
up into the bladder, it may remain in
the lower part of it, from which it
may be eafily preffed back again.
This, however, is not neceffary, fince
if it remain in the lower part of the
bladder, nothing but the pure air will

get

get into the pipe, and fo into the water.

7. If much more than half of the veffel be filled with air, there will not be a body of water fufficient to agitate, and the procefs will take up much more time.

8. If the chalk be too finely powdered, it will yield the fixed air too faft.

9. After every procefs, the water to which the chalk is put muft be changed.

10. It will be proper to fill the bladder with water once every day, after it has been ufed, that any of the oil of vitriol which may have got into it, and would be in danger of corroding it, may be thoroughly diluted.

11. The veffel, which I have generally made ufe of, holds about three pints, and the phial containing the chalk and water is one of ten ounces;

and

and I find that a little more than a tea-fpoonful of oil of vitriol is fufficient to produce as much air as will impregnate that quantity of water.

12. If the veffel containing the water be larger, the phial containing the chalk and the oil of vitriol fhould either be larger in proportion, or frefh water and oil of vitriol muft be put to the chalk, to produce the requifite quantity of air.

13. In general, the whole procefs does not take up more than about a quarter of an hour, the agitation not five minutes; and in nearly the fame time might a veffel of water, containing two or three gallons, or indeed any quantity that a perfon could well fhake, be impregnated with fixed air, if the phial containing the chalk and oil of vitriol, be larger in the fame proportion.

14. To give the water as much air

C 6

as

as it can receive in this way, the procefs may be repeated with the water
thus impregnated. I generally chufe
to do it two or three times, but very
little will be gained by repeating it of-
tener; fince, after fome time, as much
fixed air will efcape from that part of
the furface of the water which is expofed to the common air, as can be
imbibed from within the veffel.

15. All calcareous fubftances contain fixed air, and any acids may be
ufed in order to fet it loofe from
them; but chalk and oil of vitriol
are, both of them, the cheapeft, and,
upon the whole, the beft for the purpofe.

16. It may poffibly be imagined
that part of the oil of vitriol is rendered volatile in this procefs, and fo
becomes mixed with the water; but
it does not appear, by the moft rigid
chymical examination, that the leaft
perceivable

perceivable quantity of the acid gets into the water in this way; and if so small a quantity as a single drop of oil of vitriol be mixed with a pint of water (and a much greater quantity would be far from making it less wholesome) it might be discovered. The experiments which were made to ascertain this fact were made with *distilled water*, the disagreeable taste of which is not taken off, in any degree, by the mixture of fixed air. Other-wise, distilled water, being clogged with no foreign principle, will imbibe fixed air faster, and retain a greater quantity of it than other water. In the experiments that were made for this purpose, I was assisted by Mr. Hey, a surgeon in Leeds, who is well skilled in the methods of examining the properties of mineral waters.

17. Dr. Brownrigg, who made his experiments on Pyrmont water at the spring

fpring head, never found that it contained fo much as one half of an equal
bulk of air; but in this method the
water is eafily made to imbibe an
equal bulk. For it muſt be obferved,
that a confiderable quantity of the
moſt foluble part of the air is incor-
porated with the water, as it firſt
afcends through it, before it occupies
its place in the upper part of the
veſſel.

18. The heat of boiling water will
expel all the fixed air, if a phial containing this impregnated water be held
in it; but it will often require above
half an hour to effeɕt it compleatly.

19. If any perfon would chufe to
make this medicated water more nearly
to refemble genuine Pyrmont water,
Sir John Pringle informs me, that
from eight to ten drops of *Tinɕtura
Martis cum ſpiritu falis* muſt be mixed
with every pint of it. It is agreed,
however,

however, on all hands, that the peculiar virtues of Pyrmont, or any other mineral water which has the fame brifk or acidulous tafte, depend not upon its being a chalybeate, but upon the fixed air which it contains.

But water impregnated with fixed air does of itfelf diffolve iron, as the ingenious Mr. Lane has difcovered; and iron filings put to this medicated water make a ftrong and agreeable chalybeate, fimilar to fome other natural chalybeates, which hold the iron in folution by means of fixed air only, and not by means of any acid; and thefe chalybeates, I am informed, are generally the moft agreeable to the ftomach.

20. By this procefs may fixed air be given to wine, beer, and almoft any liquor whatever: and when beer is become flat or dead, it will be revived by this means; but the delicate agreeable

able flavour, or acidulous tafte communicated by the fixed air, and which
is manifeft in water, will hardly be
perceived in wine, or other liquors
which have much tafte of their own.

21. I would not interfere with the
province of the phyfician, but I cannot entirely fatisfy myfelf without
taking this opportunity to fuggeft
fuch hints as have occurred to myfelf, or my friends, with refpect to
the *medicinal ufes* of water impregnated with fixed air, and alfo of fixed
air in other applications.

In general, the difeafes in which
water impregnated with fixed air will
moft probably be ferviceable, are thofe
of a *putrid* nature, of which kind is
the *fea-fcurvy*.   It can hardly be
doubted, alfo, but that this water muft
have all the medicinal virtues of Pyrmont water, and of other mineral waters fimilar to it, whatever they be;
efpecially

especially if a few iron filings be put to it, to render it a chalybeate, like genuine Pyrmont water. It is possible, however, that in some cases it may be desirable to have the *fixed air* of Pyrmont water, without the *iron* which it contains.

Having this opportunity, I shall also hint the application of fixed air in the form of *clysters*, which occurred to me while I was attending to this subject, as what promises to be useful to correct putrefaction in the intestinal canal, and other parts of the system to which it may, by this channel, be conveyed. It has been tried once by Mr. Hey above-mentioned, and the recovery of the patient from an alarming putrid fever, when the stools were become black, hot, and very fetid, was so circumstanced, that it is not improbable but that it might be owing, in some measure, to those clysters.

clyfters. The application, however, appeared to be perfectly eafy and fafe.

I cannot help thinking that fixed air might be applied externally to good advantage in other cafes of a putrid nature, even when the whole fyftem was affected. There would be no difficulty in placing the body fo, that the greateft part of its furface fhould be expofed to this kind of air; and if a piece of putrid flefh will become firm and fweet in that fituation, as Dr. Macbride found, fome advantage, I fhould think, might be expected from the fame antifeptic application, affifted by the *vis vitæ*, operating internally, to counteract the fame putrid tendency. Some Indians, I have been informed, bury their patients, labouring under putrid difeafes, up to the chin in frefh mould, which is alfo known to take off the

fœtor

fœtor from flesh meat beginning to putrify. If this practice be of any use, may it not be owing to the fixed air imbibed by the pores of the skin in that situation? Following the plough is an old prescription for a consumption, as also is living near lime kilns. There is often some good reason for very old and long continued practices, though it is frequently a long time before it be discovered, and the *rationale* of them satisfactorily explained.

Being no physician, I run no risque by throwing out these random hints and conjectures. I shall think myself happy, if any of them should be the means of making those persons, whom they immediately concern, attend more particularly to the subject. My friend Dr. Percival has for some time past been employed in making experiments on fixed air, and he is particularly at-

tentive to the medicinal uſes of it;
and from his knowledge as a philoſo-
pher, and ſkill in his profeſſion, I
have very conſiderable expectations.

C H A P-

CHAPTER III.

OF DR. NOOTH'S OBJECTIONS TO
THE PRECEDING METHOD OF
IMPREGNATING WATER WITH
FIXED AIR, AND A COMPARISON
OF IT WITH HIS OWN METHOD,
BOTH AS PUBLISHED BY HIM-
SELF, AND AS IMPROVED BY
MR. PARKER.

I can eafily forgive Dr. Nooth for his reprefenting me as having no other merit than the *firſt publication* of the method for impregnating water with fixed air, accounting for it as I have done before; but I cannot fo eafily forgive another paragraph in his paper, the tendency of which is intirely to difcredit a method, which, though it is, in fome refpects, inferior to his own, has neverthelefs its peculiar advantages :

vantages : and every advantage cannot poffibly concur in any one method. He fays, p. 59, " Independent
" of the inconveniencies attending
" the procefs, there was another ob-
" jection to the apparatus, which,
" with moft people, might have
" confiderable weight. The *bladder*,
" which formed part of it, was
" thought to render the water offen-
" five; and when the folvent power
" of fixed air is confidered, it will not
" appear improbable, that the water
" would be always more or lefs taint-
" ed by the bladder. In fome trials
" which I made with Dr. Prieftley's
" apparatus, it always happened that
" the water acquired an *urinous fla-*
" *vour* ; and this tafte was, in gene-
" ral, fo predominant, that it could
" not be fwallowed without fome de-
" gree of reluctance."

That Dr. Nooth *did* produce an
impregnated

impregnated water which he could
not fwallow without reluctance, and
even that, in the trials to which he
refers, he *generally* produced fuch water, I am far from doubting ; becaufe
that might happen from various
caufes. But that the urinous flavour
came from the *bladder*, as fuch, I will
venture to fay is not poffible. For
then it would *always* have had the
fame effect ; and not only myfelf have
never perceived fuch a flavour as the
Doctor complains of, but this is the
only complaint of the kind that I
have hitherto heard of; though many
perfons of the moft delicate tafte, and
particularly many ladies, have ufed
the water impregnated in my method
for months together. Few perfons
have had to do with bladders, and
fixed air confined in bladders, more
than myfelf; and yet I have never
feen any reafon to fufpect this great

folvent

solvent power of fixed air with respect to them ; especially so as to be apparent in the space of a few minutes.

But supposing the fixed air to be capable of dissolving the whole bladder, and to carry it along with itself into the impregnated water, no physician, or philosopher, will pretend to say that it could have any more tendency to give it an *urinous flavour*, than if it had been any other membrane of the animal body.

Indeed, as the Doctor himself does not pretend to say that this strange urinous flavour was the effect of *all* the impregnations of water made in my method, but only in *some* of them (though it was *generally* so, in those particular trials) it is evident, from his tacit confession, that it must have been an *accidental thing*, and could not have come from the bladder, which I suppose he made use of in all trials.

For

For he has not done me the juftice to acknowledge that, in my pamphlet, among the various methods of effect-ing the impregnation of water, I have defcribed one in which no bladder is made ufe of. When the Doctor fhall once more produce this urinous fla-vour (and as a new and curious expe-riment, it is certainly worthy of his farther inveftigation) taking care that no carelefs fervant fhall have mixed any urine in the water that he calls for, I fhall give this new objection to my procefs a farther examination. At prefent I am inclined to confider this as an experiment of the fervant, ra-ther than of the Doctor himfelf.

Several perfons have thought that fixed air difcharged from *impure chalk* gives the water that is impregnated with it a difagreeable flavour, but this I have never obferved myfelf; and any other calcareous matter may be

ufed

ufed in my method, as well as in that of Dr. Nooth, who recommends chalk, as the beft upon the whole.

I fhall conclude thefe animadverfions with doing what Dr. Nooth ought to have done before me, viz. fairly ftating the advantages and difadvantages of our two methods. His method requires *lefs fkill* in the operator and *a lefs conftant attention.* It is alfo *more elegant* and cleanly, I mean with refpect to the *operator*; for this does not at all affect the *impregnated water.* On thefe accounts I generally recommend and make ufe of his method myfelf, efpecially as the glaffes are made with improvements by Mr. Parker. But if Dr. Nooth be candid, he muft acknowledge that my method requires much *lefs time,* and is much *lefs expenfive*; and therefore muft be more proper when a great quantity of impregnated water is wanted ;

ed; and especially when there is but little room to make it in.

My method indeed requires a constant attendance, but I question whether, upon the whole, more than is necessary to be given to Dr. Nooth's method at intervals, if the water be at all agitated; considering that mine does not require one-tenth part of the time. And though my method requires some little skill and address, it is not so much, but that many persons, altogether unused to experiments, have, to my knowledge, succeeded in it very well, and have made the impregnated water in a constant way for their family use, and without any assistance besides what they got from the printed directions. My apparatus costs little or nothing, because no vessels are made for the purpose; and both the chalk and the acids are made to go as far as possible, by means

of

of the convenient agitation of the veſſel in which they are contained. Whereas Dr. Nooth's method requires a pecu- liar and expenſive apparatus, and more waſte is unavoidable in the uſe of it. However, for the reaſons above- mentioned, I have never recommend- ed my own method for the uſe of a family ſince I have been acquainted with his.

What I have ſaid above is rather ap- plicable to the apparatus as it is made by Mr. Parker, than to that which Dr. Nooth has deſcribed. For Mr. Parker's glaſſes are, in my opinion, conſiderably improved from thoſe of Dr. Nooth. It may be ſaid that the improvements conſiſt in *little things* ; but little things may have great ef- fects ; and, after the diſcovery of the *firſt method* of accompliſhing this end, all *ſubſequent methods* may be called little things ; and they may be end- leſsly lefsly diverfified, without any great claim of merit. I have feen feveral very ingenious methods fince the publication of mine, though none that I liked fo much, upon the whole, as that of Dr. Nooth, improved by Mr. Parker.

In Dr. Nooth's apparatus, if any more air than is wanted be produced, the water will run out of the uppermoft veffel. To ufe his own words, p. 63, " Should more air be extricat-
" ed than is fufiicient, in the conduct
" of the procefs, to fill that veffel,
" the water will run over the top of
" it, and will continue to run as long
" as any air afcends in the middle vef-
" fel, or 'till the furface of the water
" is below the extremity of the bent
" tube; and in this cafe the whole
" would be wet and difagreeable."
But this difagreeable confequence can never happen in the ufe of Mr. Par-

ker's

ker's glaffes, becaufe the bent tube in which the uppermoft veffel terminates is made of fuch a length, that the water expelled from the middle veffel can do no more than nearly fill the uppermoft, and can never run over ; fo that whereas Dr. Nooth's apparatus requires a conftant attendance, Mr. Parker's requires none. The materials being once put into it, the procefs will go on of itfelf, without any farther care ; unlefs the operator fhould chufe to accelerate the impregnation by now and then letting out the air that is not eafily abforbed, and by agitating the water. This I think to be a confiderable advantage gained by a very eafy contrivance of Mr. Parker's, overlooked by Dr. Nooth.

Mr. Parker derives another confiderable advantage from a *channel* which he cuts in the ftopper of his
upper-

uppermoſt veſſel, or from a ſtopper
with a hole through the middle,
which Dr. Nooth has not in his ; ſo
that either the operator muſt be careful to take it out during the efferve-
ſcence, or it will be driven out, or
ſome of the veſſels will burſt, to the
great danger of the by-ſtanders; which
actually happened in one made by
Mr. Parker, before he thought of
this method to prevent it. Whereas,
through the channel in Mr. Parker's
apparatus, the common air eaſily eſ-
capes from the uppermoſt veſſel, to
make room for the water to aſcend;
and when, in the continuance of the
proceſs, the fixed air riſes through
the bent tube into the uppermoſt veſſel, it lodges upon the ſurface of the
water in it; and the communication
between it and the common air being
ſo much obſtructed, they are ſuffici-
ently ſeparated; ſo that even the wa-

D 4   ter

ter in the uppermoft veffel has (if the production of air be copious) almoft as much advantage for receiving the impregnation, as that in the middle veffel.   This advantage Dr. Nooth lofes.

Alfo, when he chufes to feparate the two uppermoft veffels from the loweft, in order to agitate the water, he muft either leave the mouth of the uppermoft veffel open, in which cafe he can hardly agitate the water at all; or (as he prefers to do it) he muft put the ftopper in, and confequently admit the common air to pafs his valve, and mix with the fixed air, which muft greatly retard the abforption of it: whereas Mr. Parker's veffels may be agitated with the ftopper in, which, admitting the common air into the upper veffel, through the channel cut in it (or through the hole of the ftopper) permits the water to
defcend

defcend into the lower, on the furface of which nothing but fixed air is incumbent. Should any common air enter by the valve, which in this cafe it hardly would, the finger of the perfon who fhakes the veffel may eafily be placed fo as to prevent it.

Laftly, I confider it as a valuable improvement in Mr. Parker's apparatus, that, by means of the openings into the middle and loweft veffels, clofed with ground ftopples, the operator is enabled to draw off his water, in order to tafte it occafionally, or to add to his oil of vitriol or chalk, &c. at pleafure, without giving himfelf the trouble of feparating the veffels from one another for thofe purpofes.

The firft apparatus that I faw of Mr. Parker's had no *valve* at all, but only a glafs ftopple, with one or more fmall perforations, for the afcent of

D 5 the

the air into the middle veffel.   This
I ftill generally make ufe of, without
finding any occafion for a valve; the
afcent of the fixed air fufficiently preventing the defcent of the water, as
long as the procefs continues, efpecially when pounded *marble* is ufed.
This fubftance Dr. Franklin recommended to me, and I give it the pre-
ference very greatly to chalk, chiefly
on account of the length of time that
is required to expel the air from it:
For without any frefh acid, it will
often continue to yield air for feveral
days together.

That thofe perfons who are not
poffeffed of the Englifh *Philofophical
Tranfactions*, and particularly foreign-
ers, may underftand what has preceded, I fhall give a drawing of Dr.
Nooth's apparatus *, as improved by

---

* Fig. 4.

Mr.

Mr. Parker, with the following general defcription of it.

In the loweft veffel, the chalk or marble, and the water acidulated with oil of vitriol, muft be put, and into the middle veffel the water to be impregnated. During the effervefcence, the fixed air rifes into the middle veffel, and refts upon the furface of the water in it, while the water that is difplaced by the air rifes through the bent tube into the uppermoft veffel, the common air going out through the channel in the ftopple. When the bent tube is of a proper length, the procefs requires no attention; and if the production of air be copious, the water will generally be fufficiently impregnated in five or fix hours. At leaft, all the attention that needs be given to it is to raife the uppermoft veffel once or twice, to let out that part of the fixed air which is not

readily

readily abforbed by water. If the operator chufe to accelerate the pro- cefs, by agitating the water, he muft feparate the two uppermoft veffels from the loweft. For if he fhould agitate them all together, he will oc- cafion too copious a production of air; and he will alfo be in danger of throwing the liquor contained in the loweft veffel into contact with the ftopple which feparates it from the middle veffel, by which means fome of the oil of vitriol might get into the water.

End of the Extract from Dr. Priest-ley's Experiments on Air, Vol. II.

APPEN-

APPENDIX.

DR. NOOTH'S METHOD OF IMPREG-
NATING WATER WITH FIXED
AIR, AS IMPROVED BY MR. PAR-
KER, MR. MAGELLAN, &c.

Defcription of the Apparatus.

See Fig. 4.

IT is made of glafs, and ftands on a
wooden veffel *d d* refembling a
tea-board, to catch any water that
may chance to be fpilled, and prevent
it from falling on the table. The
middle veffel B has a neck which is
inferted into the mouth of the veffel
A, to which it is ground air-tight.
This lower neck of the veffel B, has a
glafs ftopple S, compofed of two
parts, both having holes fufficient to

let

let a good quantity of air pafs through them. Between thefe two parts (which may be confidered as two ftopples) is left a fmall fpace, con-taining a plano convex lens, (that is, a glafs round on one fide and flat on the other) which acts like a valve, in letting the air pafs from below up-wards, and hindering its return into the veffel A.

The upper veffel C terminates be-low in a tube *r t*, which being crook-ed, hinders the immediate afcent of the bubbles of fixed air into that vef-fel, before thcy reach the furface of the water in the veffel B. The veffel C is alfo ground air-tight to the upper neck of the middle veffel B, and has a ftopple *p* fitted to its upper mouth, which has an hole through its mid-dle. The upper veffel C holds juft half as much as the middle one B; and the end *t* of thc crooked tube,

goes

4

goes no lower than the middle of the vessel B.

The Process.

Fill the middle vessel B with spring, or any other clean and wholesome water, and join to it again the upper vessel C. Pour water into the vessel A (by the opening *m*, or otherwise) so as to cover the rising part of its bottom. About three quarters of a pint, or a little more, will be sufficient. Fill an ounce phial with oil of vitriol, and add it to the water, shaking the vessel so as to mix them well together. As heat is generated, it will be better to add the oil by a little at a time, otherwise a hazard is run of breaking the vessel. Put to this, through a wide glass, or paper funnel, about an ounce of powdered raw chalk, or marble*. The

funnel

---

* White marble being first granulated, or pounded like coarse sand, is much better for the purpose

funnel muſt be uſed in order to prevent the powder from touching the inſide of the veſſel's mouth ; for if that happen, it will ſtick ſo ſtrongly to the neck of the veſſel B, as not to admit of their being ſeparated without breaking. Place immediately the two veſ-ſels B and C (faſtened to each other) into the mouth of the veſſel A, as in the figure, and all the fixed air which is diſengaged from the chalk or mar-

---

purpoſe than pounded chalk, becauſe it is harder ; and therefore the action of the diluted acid upon it is ſlower, and laſts a very conſiderable time. The ſupply of fixed air from it is therefore much more regular than with the chalk. In general, it continues to furniſh fixed air more than twenty-four hours. When no more air is produced, if the water be decanted from the veſſel A, and the white ſediment waſhed off, the remaining granulated marble may be employed again by adding to it freſh water, and a new quantity of oil of vitriol. A farther produce of fixed air will then be furniſhed, and this may be repeated until all the marble be diſſolved.

ble

ble by the oil of vitriol, will pafs up through the valve in S into the veffel B. When this fixed air comes to the top of the veffel B, it will diflodge from thence as much water as is equal to its bulk; which water will be forced up through the crooked tube into the upper veffel C.

Care muft be taken not to fhake the veffel A when the powdered chalk is put in; otherwife a great and fudden effervefcence will enfue, which will perhaps expel part of the contents. In fuch cafe it may be neceffary to open a little the ftopple *m*, in order to give vent, otherwife the veffel A may burft. It will be proper alfo to throw away the contents, and wafh the veffel; for the matter will ftick between the necks of the veffels, and cement them together. The operation muft then be begun afrefh. But if the chalk be thrown in without fhaking .

fhaking the machine, or if marble be
ufed, the effervefcence will not be
violent. If the chalk be put into the
veffel loofely wrapt up in paper, this
accident will be ftill better guarded
againft. When the effervefcence goes
on well, the veffel C will foon be filled
with water, and the veffel B half
filled with air; which will eafily be
known to be the cafe by the air going
up in large bubbles through the
crooked tube *r t*.

When this is obferved, take off the
two veffels B and C together as they
are, and fhake them fo that the water
and air within them may be much
agitated. A great part of the fixed
air will be abforbed into the water;
as will appear by the end of the crooked tube being confiderably under the
furface of the water in the veffel.
The fhaking them for two or three
minutes will be fufficient for this purpofe.

pore. Thefe veffels muft not be fhook
while joined to the under one A,
otherwife too great an effervefcence
will be occafioned in the latter; together with the ill confequences above-
mentioned. After the water and air
have been fufficiently agitated, loofen
the upper veffel C, fo that the remaining water may fall down into B, and
the unabforbed air pafs out. Put
thefe veffels together, and replace
them into the mouth of A, in order
that B may be again half filled with
fixed air. Shake the veffels B and C,
and let out the unabforbed air, as before. By repeating the operation three
or four times, the water will be fufficiently impregnated.

Whenever the effervefcence nearly
ceafes in the veffel A, it may be renewed by giving it a gentle fhake, fo
that the powdered chalk or marble at
the bottom may be mixed with the oil
of

of vitriol and water above it; for then a greater quantity of fixed air will be disengaged.

When the effervescence can be no longer renewed by shaking the vessel A, either more chalk must be put in, or more oil of vitriol; or more water, if neither of these produce the desired effects.

The ingenious Mr. Magellan has still farther improved the contrivance of Dr. Nooth and Mr. Parker. He has two sets of the vessels B and C. While he is shaking the air and water contained in one of these sets, the other may be receiving fixed air from the vessel A. By this means twice the quantity of water may be impregnated in the same time. He has a wooden stand K (Fig. 5.) to fix the vessels B C on, when taken off from A, which is very convenient. He has a small tin trough for measuring the quantity

quantity of chalk or marble requisite for one operation, and a wide glass funnel for putting it through into the vessel A, to prevent its sticking to the sides, as mentioned before.

He has also contrived a stopple without an hole to be used occasionally instead of the perforated one *p*. It has a kind of bason at the top to hold an additional weight when necessary. (See Fig. 6.) The stopple must be of a conical figure, and very loose; but so exactly and smoothly ground as to be air-tight merely by its pressure, which may be encreased by additional weights put into its bason. Its use is to compress the fixed air on the water, and thereby encrease the impregnation. For by keeping the air on the water in this compressed state, the latter may be made to sparkle like Champaign. And if the vessels be

strong,

ftrong, there will be no danger of their burfting in the operation.

If the veffels be fuffered to ftand fix or eight hours, the water will be fufficiently impregnated even without agitation. But by employing the means above defcribed, it may be done in as many minutes.

The water thus impregnated may be drawn out at the opening *k*. But if it be not wanted immediately, it will be better to let it remain in the machine, where it has no communication with the external air. Otherwife the fixed air flies off by degrees, and the water becomes vapid and flat; as alfo happens to other acidulous waters. But it may be kept a long time in bottles well ftopped, efpecially if they be placed with their mouths downwards.

This water is more pleafant to the tafte than the natural Pyrmont or

Seltzer

Seltzer waters; as, befides their fixed air they contain faline particles of a difagreeable tafte, which are known to contribute little or nothing to their medicinal virtues; and may, in fome cafes, be hurtful. The artificial wa-ter *is* alfo double the ftrength of the natural; the latter containing fcarce half of the fixed air which can thus be communicated to the former.

N. B. Mr. Blades, of Ludgate-Hill, has ftill further improved this apparatus, by changing the ftopple at *k* for a glafs cock, which is more convenient. He has likewife altered the middle veffel B into a form more advantageous for the impregnation. See Fig. 7. With it are alfo given, a phial for meafuring the vitriolic acid, a tin meafure for the chalk or marble, and a glafs funnel to pafs it through.

A METHOD

A METHOD OF IMITATING THE
SULPHUREOUS MINERAL WA-
TERS, BY IMPREGNATING WA-
TER WITH HEPATIC AIR.

We may imitate the *sulphureous* mineral waters, as well as the *acidulous*
ones, or thofe impregnated with fixed
air. The procefs is fufficiently fim-
ple; and the fame apparatus will
ferve for both.

Inftead of limeftone, chalk, or
marble, *liver of fulphur* is to be ufed.
It may be bought ready prepared of
the chymifts or apothecaries; or may
eafily be prepared as follows:

Mix together equal parts of brim-
ftone, and of clean pot afhes*, and
place them in a crucible, or unglazed
difh, over a very gentle fire. Keep
them ftirring with a ftick 'till they

* Quick lime may be ufed inftead of pot afhes,
taking care to chufe it well burnt.

are

are united together into a blood-red mafs. Put it, while warm, into a bottle, which is to be kept well clofed.

Put a fufficient quantity of this fubflance, with the oil of vitriol and water, into the part A of the apparatus, and proceed as defcribed in the procefs for impregnating water with fixed air; the *hepatic* air will arife; the water in the middle veffel B will be impregnated with it, will fmell ftrongly fulphureous, and refemble the celebrated waters of *Aix la Chapelle*, &c. in the fame manner as thofe impregnated with fixed air refemble thofe of *Pyrmont* and *Seltzer*.

The water thus impregnated may be heated, by putting it into a clofe veffel, placed in one that contains boiling water, and it is then a *warm fulphureous water.*

If it be not ufed immediately, it

E   fhould

fhould be preferved in glafs or ftone bottles, well corked, and cemented, and placed with the corks downward in a cellar.

To imitate more exactly the feveral Mineral Waters.

This confifts only in adding to the water to be impregnated, the folid matters which they are found to leave behind on evaporation. For example:

I. PYRMONT WATER.

Add to the water in the middle veffel B, in the proportion of about 30 grains of vitriolated magnefia*, ten grains of common falt, two fcruples of magnefia alba, two fcruples of chalk, and a dram of iron filings, or

---

* Epfom falt, or Sal Catharticus Amarus. In this edition the names of the New London Pharmacopœia are commonly ufed.

iron

iron wire, clean and free from ruſt, to one gallon of water, and impregnate the whole with fixed air in the manner deſcribed. Let them remain 'till the other ingredients, and as much of the iron as is neceſſary, are diſſolved, which will be in two or three days.

2. SPA WATER.

Take of natron and magneſia of each a ſcruple, of common ſalt eight grains, water a gallon; impregnate them with fixed air; a few iron filings muſt alſo be added.

3. SELTZER WATER.

Take of natron ſeven ſcruples, common ſalt a dram and half, magneſia one ſcruple, water a gallon, and impregnate them with fixed air.

4. SEIDSCUTZ PURGING WATER, (reſembling our EPSOM.)

Take of vitriolated magneſia three

ounces, water a gallon, and impregnate them with fixed air.

5. AIX-LA-CHAPELLE WATER.

Take of fea falt two fcruples, natron a dram and half, chalk two fcruples, water a gallon. Impregnate them with hepatic air, after having firft caufed them to abforb ninety-fix ounce meafures of fixed air.

Other waters may in like manner be imitated by adding Epfom falt for purging waters, fea falt for falt waters, &c. And as fome waters (as the cold fulphureous ones) contain both *fixed* and *fulphureous* air, a mixture of liver of fulphur and chalk may be put into the veffel A with the oil of vitriol, by which means both thefe airs will be produced, and the water of courfe impregnated with them. In making artificial mineral waters, diftilled water ought always to be ufed.

AN

ACCOUNT

OF THE

NATURE, PROPERTIES,

AND

MEDICINAL VIRTUES

OF THE

Principal Mineral Waters

IN

GREAT BRITAIN AND IRELAND;

AND OF THOSE

MOST IN REPUTE IN FOREIGN PARTS.

Digested into Alphabetical Order.

By JOHN ELLIOT, M. D.

E 3

INTRODUCTION.

THE following treatife on mineral waters being intended for the Public in general, the Editor has endeavoured to couch it in fuch terms as that it may be underftood by thofe who are unacquainted with the art of phyfic. Such an account has been judged by many very proper to be fubjoined to the foregoing differtation.

All the mineral waters in *England,* of any note, will be found noticed in this tract: together with their virtues, and the method, and feafon of ufing them, fo far as could be learnt from the authors who have been confulted on the occafion. To thefe are added all the principal mineral waters of *Scotland* and *Ireland,* as well as the moft celebrated ones which the English

glifh

glifh valetudinarian may have occafion to vifit on the continent.

The greateft part of the books
which have hitherto been written on
this fubject, abound with experiments
tending to fhow the *analyfis* of thofe
waters. But this can be of little ufe
except to the faculty; and muft be
dry, and perfectly unintcrefting to
common readers. Befides, the necef-
fity of fuch accounts is fuperceded by
fpecifying the ingredients themfelves
with which the waters are impregnated, and their virtues as medicines;
to fhow which is the fole end of thefe
experiments. It would alfo have
fwelled the volume to an unwieldy
fize. For this laft reafon, as alfo becaufe it was judged wholly unnecef-
fary and fuperfluous, the defcriptions
of the places in which the refpective
waters are fituated, are likewife omitted.

For

For the convenience of the reader, the waters are arranged in *alphabetical* order, by which means they will the more readily be found. I wonder indeed that this method is not obferved by authors on many other occafions. For though there be a fyftematical arrangement of the things treated of in their books, yet the reader is, after all, obliged to refer to an *index*; which in fact is an alphabetical arrangement of the particulars of the fubject.

The reader will find accounts of a great number of waters which he probably never heard of before. As many of thefe are of fimilar virtues to others which are more famous, the invalid will be inftructed where to find a mineral water proper for his complaint near at hand, when it might not be convenient for him, on account of the diftance, or otherwife, to repair to thofe of greater *note*, though perhaps not of fuperior *virtue*.

For this purpofe alfo, as well as for the more readily finding out waters of particular virtues, the waters are alfo claffed or arranged according to their refpective mineral properties; as will prefently be feen.

Water, from the nature of the foil over which it paffes, and other accidents to which it is expofed, is always more or lefs impregnated with foreign particles. According to the nature of thefe particles, the properties of the water are different. Hence we have *hard* water, *foft* water, *falt* water, and the almoft infinite variety of *mineral* waters. The principal of the latter, in this part of the world, will be found noticed in the following tables.

1ft. CHA-

1ft. CHALYBEATE WATERS.

Hampftead	Glendy
Carlton	Aberbrothick
Iflington	Cobham
Leez	Tunbridge
Markfhall	Buxton
Felftead	Millar's Spa
Wellenbrow	Latham
Aylefham	Tibfhelf
Malvern	Chippenham
Colurian	Witham
Harrogate	Lancafter
Road	Whiteacre
Ilmington	Weft Afhton
Birmingham	Cawthorp
Cannock	Derby
Mofs Houfe	Weatherftack
Wigan	Filah
Sene	Dortfhill
Thetford	Stanger
Lincomb	Dunfe
Llandrindod	Caftle Connel
Peterhead	Tralee

Granfhaw

Granſhaw	Wexford
Newtown Stewart	Ballyſpellan
Galway	Nezdenice
Coolauran	Kuka
Liſdonvarna	Spa
Ballycaſtle	Zahorovice
Glanmile	Bromley
Kanturk	* Bath
Dunnard	* Matlock.
Maccroomp	

Chalybeate waters are the moſt uſeful and beneficial to health of any of the mineral waters; and are very plentiful in this iſland.

Waters are known to be chalybeate by their ſtriking a reddiſh purple, or black colour with an infuſion of galls; and according to the height of the colour, provided the ſtrength of the infuſion be the ſame, we judge of the ſtrength of the water as a chalybeate.

The iron in thoſe waters is held in ſolution by means of fixed air, as may be judged from what has been already

7 ſaid

faid on this fubject. As the fixed air
foon flies off on expofing the water,
the iron falls to the bottom in form
of a brownifh yellow powder. Hence
thefe waters ftrike the' deepeft black
with galls at the fpring head ; and in
time they wholly lofe that property.

They have a brifk acidulous or vinous tafte when frefh, and tinge the
ftools black.

Taken inwardly they ftrengthen
the conftitution in general, increafe
the tone of the fibres, quicken the
circulation, and reftore a proper confiftence to the blood when in a too
thin and watery ftate. And hence
they are found to invigorate the
whole frame. They are good in
difeafes arifing from weaknefs; in
fpafmodic diforders, arifing from too
great irritability and relaxation of the
nervous fyftem ; in fluor albus, and
gleets ; in female obftructions ; in
hyfteric

hyfteric and hypochondriacal diforders; in lofs of appetite and digeftion;
and in a variety of other complaints,
as will be fpecified in treating of the
refpective waters; they differing fomewhat in their virtues.

Previous to a courfe of thefe waters, bleeding, and a cooling purge,
may be neceffary, in cafe of heat and
fever; and coftivenefs fhould alfo be
avoided while drinking them. Where
there is much fever, and alfo in ulcers
of the lungs, and in confirmed obftructions attended with fever, the ufe
of thefe waters is improper.

Patients ought to begin with drinking a fmall quantity of thefe waters
every morning, and gradually to increafe the dofe. A temperate and
moderate diet, and gentle exercife
fhould alfo be obferved while taking
them.

If the water be too cold for the
ftomach,

ftomach, a bottle containing fome of it may be placed in warm water juft before drinking.

Acids, tea, and other things, which decompofe thefe waters, fhould not be taken for fome time before or after drinking them.

Befides iron, thefe waters ufually contain fea falt, natron, a purging falt, or other fubftance, as will be noticed when treating of them.

2d. CHALYBEATE PURGING WA-TERS.

Knowfley	Thirfk
Burlington	Hartlepool
Aftrope	Thornton
Coventry	Orfton
Bournley	Stenfield
Townley	Kirby
Newham Regis	Tarleton
Binley	Malton
Kingfcliff	Afwarby
	Scarborough

Scarborough	Egra
Cheltenham	Nevil Holt
Bagnigge	Ballycaftle
Stoke	Deddington
Woodham Ferris	Drig-Well
Hanlys	Inglewhite
Athlone	Gainfborough
Mount Pallas	Thorp Arch
Killinfhanvally	Caftlemaign
Cleves	Ballynahinch
Hoff Geifmar	Jeffop
Pyrmont ·	Driburg.

Thefe chalybeate waters contain a greater proportion of purging falt than of any other folid matter, and therefore when taken in fufficient quantity (feveral pints) they operate by ftool. They have this advantage over other purges, that they do not exhauft the ftrength.

If taken in lefs quantity, as alteratives, they operate chiefly by urine,

and

and then they fall rather under the firſt claſs of theſe waters than the preſent.—*See what was ſaid of chaly-beate waters.*

3d. SULPHUREOUS WATERS.

Sutton Bog	Hariogate
Wigleſworth	Maudſby
Chadlington	Crickleſpaw
Bilton	Broughton
Queen Camel	Shettlewood
Nottington	Reddleſtone
Drumgoon	Durham
Swadlingbar	Wardrow
Derryleſter	Skipton
Liſbeak	Landrindod
Killaſher	Moffat
Mechan	Corſtorphin
Aſhwood	Caſtle Loed
Derryhence	Fairburn
Drumaſnave	Rippon
Anaduff	Groſſenendorf
Aphaloo	* Aix la Chapelle
	* Borſet

* Borfet   * Baden Baden

* Bareges   * Saint Amand.

Waters called *fulphureous* do not contain an actual fulphur, but are impregnated with a gas, or fpirit (the hepatic air already defcribed) which gives them their fulphureous fmell. Befides this, they ufually contain either natron, fea falt, a purging falt, iron, an earth, or other matter, and commonly feveral of thefe in different proportions.

Waters of this fort are diuretic, and ftrongly diaphoretic, and are therefore good in cutaneous difeafes, ufed both internally and externally. They are alfo good in chronic obftructions; and in diforders proceeding from acidity, from worms, &c.

They ufually make filver appear of a copper colour.

4th, SUL-

4th. SULPHUREOUS PURGING WATERS*.

Afkeron	Upminfter
Croft	Codfalwood
Cawley	Wirkfworth
Cunley Houfe	Derrindaff
Buglawton	Owen Bruen
Loanfbury	Pettigoe
Normanby	Enghien
Shapmoor	

Thefe waters differ from thofe in the preceding clafs in containing a purging falt as the principal folid ingredient, and therefore operating by ftool. They are good in the fame diforders as the alterative fulphureous waters, as alfo for foulneffes of the bowels, &c.

---

* Some of the chalybeate purging waters are alfo fulphureous.

5th. ACI-

5th. ACIDULOUS, OR SALINE WA-
TERS.

Seltzer	Cape Clare
Tilbury	Buch
Clifton	Tonſtein
Glaſtonbury	* Mount d'Or
Toberbony	* Chaude Fontaine
Carrickmore	* Piſa.
St. Bartholomew	

The waters of this claſs contain natron. This ſalt, as the waters are taken up from the fountain, is ſaturated, or rather ſuperſaturated, with fixed air; hence the waters do not then manifeſt any alkaline quality; on the contrary, they curdle with ſoap, and are termed *acidulæ*. This *fixed air*, or *aërial acid*, however, being very volatile, ſoon exhales when the water is heated, or ſtands awhile expoſed, and then the alkali manifeſts itſelf.

The

The general virtues of thefe waters may be known from what is faid in the alphabet, under the article SELT-ZER WATER.

6. SALINE PURGING WATERS.

Barrowdale	Acton
Leamington	Epfom
New Cartmal, or	Alkerton
Rougham	Ball, or Bandwell
St. Erafmus	Llandrindod
Cargyrle	Kenfington
Dortfhill	Richmond
Alford	Upminfter
Dulwich	Seidlitz
Holt	* Balaruc
Stretham	Sea Water
Kilburn	Dog and Duck
Moreton-fee	Kinalton
Hanlys	Brentwood
Conmer	Colchefter
Bagnigge	Sydenham
Barnet	Carrickfergus
North-hall	* Bagniers.

Thefe

These waters are impregnated with sea salt, and also with a purging salt. This, which has formerly received various names from different authors, is now generally suppofed to be *vitriolated magnesia:* though, from difference of figure and solubility, many incline to think, that there may be different purging salts in different, or even in the fame waters. They who hold the former opinion, attribute this diversity of appearance to a combination with different ingredients.

They differ in strength; fome of them purge sufficiently in the quantity of a pint; while two, three, four, five, or fix pints of others are necessary to produce that effect. Some again are fo weak as to require the addition of fome other purgative falt.

Given in small quantities they act as diuretics and alteratives.

They

They are good in fcrophulous and fcorbutic complaints, ulcers, and other difeafes which make their appearance on the fkin, and are likewife ufed as baths, and fomentations in thefe and other diforders.

The virtues of the preceding clafs of waters depend in a great meafure on the prefence of their *fixed air*. The waters of the prefent clafs feem to derive their virtues principally from the faline matters which they contain.

7. VITRIOLIC WATERS.

Shadwell	Hartfel
Weftwood	Crofs-town
Swanzy	Nobber
Haigh	Cafhmore
Vahls	Kilbrew.

Thefe waters are impregnated with green vitriol or copperas, and ftrike a black colour with galls.

They are chiefly ufed outwardly for

for wafhing old fores and the like, and frequently with good effect. In fome cafes, however, they are taken inwardly in fmall dofes, and then they prove emetic and purgative.

8. WATERS WHICH CONTAIN AN EARTH.

Newton-dale	* Briftol
Bale	* Buxton
Knarefborough	* Mallow.
Glavely	

The cold waters of this clafs have a petrifying quality. The virtues of the waters of this clafs being different, the reader is referred to the refpective articles in the alphabet for an account of them.

The above arrangement of mineral waters is intended more for the convenience of the reader not verfed in phyfic, than as a *fyftematical* one.

Had the latter idea been adopted, it would

would have been neceſſary perhaps to have made a diviſion of the waters into *hot* and *cold*, in imitation of the learned Dr. Donald Monro; from whoſe ingenious work, together with thoſe of Dr. Short, Dr. Rutty, and a few others, the following treatiſe has been chiefly compiled*.

There are a great number of *cold* mineral waters in England; but the number of the *hot* ones is very ſmall. In the above catalogue, the *latter* are diſtinguiſhed from the *former* by having an ASTERISK placed before them. Thoſe of greateſt note on the continent, however, are alſo noticed; in many parts of which they abound.

The cauſe of the heat of thoſe waters is, in ſome inſtances, ſubterraneous fire; as is the caſe with ſome

* The quantity of waters to be drank, and ſome other particulars, are not always mentioned by authors, but they may eaſily be learnt on the ſpot.

F   which

which are fituated near volcanos. In
other cafes the heat arifes from the
mineral ingredients with which they
are impregnated in their paffage.
And the fame may be faid of thofe
waters which are *cooler* than the common temperature of the atmofphere.
Thus it is known, that quick-lime,
the pyrites ftone, and other fubftances, thrown into water will make
it *warm*. On the contrary, falts of
various kinds make it *colder* than before.

The *warm* waters poffefs many of
the virtues and properties of *cold* waters of the fame clafs, and which are
impregnated in the fame manner ; but
they are preferable in many cafes,
as from their warmth they are more
kindly and agreeable to the ftomachs
of weak people, and promote perfpiration.

The warm waters are alfo ufed as

warm

warm baths, and may in general be considered as warm medicated baths; and these by relaxing the fibres, are of use in a variety of disorders which take their rise from rigidity, and from spasm, as also from other causes. Hence their great use in rheumatisms, inflammations, costivenefs, &c. The cure is usually assisted by the internal use of those waters at the time.

For complaints of a particular part of the body, either the part is foment-ed with the warm water, or the water is raised to an height by pumps, or otherwise, and then let fall with force on the difeafed part; this is called *pumping* by the English; the French term it the *Douche*.

Contrivances are also used for raising these waters into *vapour* or *steam*, and confining it so that it may be applied to the whole body, or to particular parts. Thefe contrivances are called *vapour baths*.

Baths are likewife made of the mud found at the bottom of thefe waters; and they have been found ferviceable in removing pains, and achs; and paralytic, and other complaints. The mud is either rubbed on the part, or the part is immerfed in it, as may be judged convenient or proper; when it is collected in quantity in a refervoir for thefe purpofes, it is called the *mud bath*.

The cold waters are alfo, in fome cafes, ufed externally.

I fhall conclude this introduction by mentioning fome of the moft obvious methods of analyzing, or difcovering the nature of mineral waters.

The various fubftances occafionally found united with water, and with each other by diffufion, or by chemical

cal

cal folution, may be comprifed chiefly, as Dr. Fothergill obferves, under four claffes.

1. AERIAL. Atmofpheric, vital, fixed, inflammable, hepatic, and phlogifticated airs.

2. SALINE. Vitriolic, nitrous, and marine acids; natron, kali, ammonia, and fulphurated kali.

3. METALLIC. Iron, copper, zinc, manganefe, arfenic.

4. EARTHY. Magnefia, lime, clay, barytes, filiceous earth.

Of neutral falts we find the vitriolic acid united with natron, kali, lime, magnefia, clay, iron, copper, and zinc. The nitrous acid with the four former of thefe. The marine acid with the fame; and fometimes with barytes, manganefe, and clay. And the aerial acid with thefe, and

alfo

alſo with iron, zinc, and manga-
neſe.

Sulphur, foſſil oil, and extracts from
vegetable and animal ſubſtances, are
alſo found ſometimes in mineral waters.

From theſe all the virtues of mineral waters are derived, if we except
what they obtain from their temperature. To inveſtigate them by an accurate analyſis, ſome care and attention are neceſſary. The following
methods are collected from the beſt
writers on that ſubject.

Previous to the chemical examination the ſenſible qualities of the water, as taſte, ſmell, colour, and degree
of tranſparency, ſhould be obſerved.
Theſe, with the ſpecific gravity, temperature, and ſurrounding ſoil, will
afford conſiderable information, and
point out the readieſt methods of analyzing it.

To

To the *tafle* the aerial acid gives a gentle fmartnefs or poignancy : vitriolic or nitrous falts, a bitternefs : lime or felenite, a flight aufterity : alum, a fweetifh aftringency : natron, and marine falt, a naufeous brackifhnefs : copper, a flight tafte of brafs : iron, an inky tafte.

To the *fmell* aerial acid exhibits an agreeable penetrating odour like that of fermenting liquors : hepatic air *, an odour like that of a foul gun, or ignited gunpowder.

A brown, reddifh, or yellow *colour*, betrays various impurities : a whitifh indicates clay : a blue, vitriol of copper : a green or variegated film, vitriol of iron ; and this laft is confirmed if there be a yellow ochry fediment.

The examination ought to be made

---

* A bituminous or afphaltic air gives a fmell fomewhat fimilar to this.

in the different feafons, at different
times of the day, and particularly in
different ftates of the atmofphere, as
thefe have confiderable influence on
waters.

———————————

There are three modes of analyzing
mineral waters : by REAGENTS; by
EVAPORATION; by DISTILLATION.
All thefe have their ufes.

A great number of different reagents have been employed, of which
the following are the principal, and
perhaps all that deferve to be noticed.

SYRUP OF VIOLETS. All vegetable blues turn red with acids; green,
with alkalis. This has been moft
commonly ufed, but many are now
difpofed to rejeƈt it in favour of others.
It will fometimes change green with
iron ; which, if it were trufted to
alone, might lead to miftakes.

TINC-

TINCTURE OF TURNSOLE, or a blue tincture prepared from lacmofs, appears to be a more fenfible teft; and

The JUICE OF RED CABBAGE, recommended by Mr. Watt, may be in fome cafes preferable to either.

INFUSION OF BRAZIL WOOD, which is red, with alkalis becomes blue. Acids change it yellow, and reftore the red deftroyed by an alkali. Paper ftained with the infufion, a little ftarch being previoufly boiled in it, is a more fenfible teft than the infufion itfelf.

INFUSION OF TURMERIC is made brown by alkalis.

TINCTURE OF GALLS in fpirit of wine. This fhould be made as ftrong as poffible. It readily difcovers iron, in proportion to the quantity of which it will vary in colour through different gradations of purple, and if the quan-

F 5 tity

tity be large it will appear quite black.

PHLOGISTICATED, or as it is now more generally called, PRUSSIAN AL-KALI, is alfo ufed as a teft of iron, with which it exhibits Pruffian blue. It precipitates copper of a reddifh brown colour; zinc and manganefe, white; but thefe two precipitates may be diftinguifhed from each other by the latter becoming black by calcination, which effects no change in the former: it likewife precipitates other metals. An improved method of preparing this alkali may be found in the firft volume of the Analytical Review.

CONCENTRATED VITRIOLIC A-CID. It difcovers barytes.

FUMING NITROUS ACID is recommended by Bergman to precipitate fulphur, when the water contains it in the form of hepar.

ACID OF SUGAR is a very fenfible
                                            teft

teft of lime, but does not always detect it, being incapable of difengaging
it when held in folution by a confiderable excefs of any acid, the fparry
and acetous excepted. This is a curious fact not generally known.

FIXED VEGETABLE ALKALI precipitates all earths, except barytes and
metals. M. de Fourcroy recommends
it to be perfectly pure, or cauftic; but
obferves, that it will in that ftate precipitate any lefs foluble neutral falt
with an alkaline bafe.

VOLATILE ALKALI. This, if
perfectly pure, decompofes only earthy
falts with bafes of clay or magnefia;
but if aerated, will alfo decompofe calcareous falts by double affinity. It
changes water containing copper
blue.

LIME WATER detects the prefence
of aerial acid, with which it forms a
precipitate. As thirty-two parts of

F 6　　　　chalk

chalk contain thirteen of the aerial acid, the quantity of the latter, in a mineral water, may be afcertained by the weight of the chalk depofited. It alfo decompofes metallic falts, and clay or magnefia when united with the marine or vitriolic acids.

SALITED BARYTES is a moft fenfible teft of vitriolic acid, taking it from every other bafe, and forming with it an infoluble compound.

NITRATED SILVER, when diffolved in diftilled water, will detect the fmalleft veftige of a marine acid: but it is by no means an accurate teft, as vitriolic acid, if in confiderable quantity, occafions alfo a precipitate with it; and the fame effect is ftill more evidently produced by fixed alkali, chalk, or magnefia, unlefs nitrous acid fufficient to faturate them be previoufly added.

NITRATED MERCURY. Of this

there

there are two kinds, one made with heat, the other without. Many cir-cumftances combine to render this an extremely fallacious teft.

A SOLUTION OF ARSENIC in the marine acid will precipitate fulphur from water in which it is held dif-folved by means of fixed air.

We may add, that WHITE ARSE-NIC becomes yellow if immerfed in water containing hepatic gas : and a piece of polifhed iron will receive a copper-colour from water in which copper is diffolved. By the latter method copper has been detected in pine-apple rum, in which the aqua ammoniæ produced no change.

The examination of mineral wa-ters has generally been made with too fmall quantities. The beft method is to mix feveral pounds with each reagent, till the latter ceafes to pro-duce

duce any precipitate. It fhould then be fuffered to fubfide for twenty-four hours in a well-covered veffel, and, after being filtered, the precipitate may be weighed and examined.

EVAPORATION is the fecond means emp'oyed for obtaining the fixed prin-ciples of a mineral water. For this purpofe a large quantity fhould be employed; fometimes even feveral hundred pounds. Veffels of metal fhould by no means be ufed. The beft methods are, evaporating to dry-nefs in open glafs veffels in the water-bath, or, which is preferable, in glafs retorts in a fand-bath.

The refiduum thus obtained is to be weighed, and put into a phial with three or four times its weight of fpirit of wine. The phial being well fhaken, it fhould be fet by for fome hours to fubfide. What the fpirit will

will not diffolve, being dried, fhould be mixed with eight times its weight of cold diftilled water, weighing it again previoufly to afcertain the quantity taken up by the fpirit. What is not foluble in this proportion of cold water, fhould be boiled in four or five hundred times its weight of theifame fluid. All thefe products, with the refiduum of the latter, are to be examined feparately.

The fpirituous folution will contain calcareous and magnefian muriate. After evaporating to drynefs, the refiduum is to be diffolved in water. Add to this vitriolic acid: the calcareous earth will precipitate in the form of felenite, and the magnefian may be obtained in that of Epfom falt, from which kali will precipitate the magnefia.

The cold water will have diffolved the neutral falts with alkaline or earthy bafes,

bafes, and fometimes a fmall quantity of martial vitriol. As a greater or lefs number of thefe are almoſt always mixed, and in various proportions, fome care is neceſſary to afcertain them. They ſhould be feparated, if practicable, by a flow evaporation. In this way they make their appearance according to their promptitude to cryſtallize. But as this method does not always fucceed perfectly, however careful we may be in conducting the evaporation, it will be neceſſary to re-examine the falts obtained at the different periods of the procefs. Alkaline falt is known by its lixivious taſte and effervefcence with acids. Diſtilled vinegar will determine whether this be vegetable or mineral, as with the former it yields a deliquefcent falt; with the latter, foliated cryſtals. Neutral falts compofed of vitriolic acid may be decompofed by

the

the falited barytes. The vitriolic acid
will decompofe thofe into which the
nitrous or marine acid enters : if it be
the former, red fumes will arife ; if
the latter, grey. The bafes of falts
compounded of the vitriolic acid may
be diftinguifhed by the figure of the
cryftals, except natron and magnefia;
but the latter renders lime-water turbid, the former does not. If the acid
be the marine, acid of tartar will take
from it kali, and a true tartar will
be precipitated : if it be united with
natron, no decompofition will enfue.
The vitriolic acid will take from it
calcareous earth, and form felenite;
or magnefia, and form Epfom falt;
or clay, and form alum. If copper be
the bafis, aqua ammoniæ will render
the folution blue ; if iron, tincture of
galls will turn it purple or black.
Cretaceous foda, if there be any, is
ufually depofited with the muriatic
falts :

-falts: they may be feparated, how-ever, by the following procefs of M. Gioanetti. Wafh the mixed falt with diftilled vinegar: dry it and pour on fpirit of wine: this will dif-folve the acetous foda, without act-ing on the marine falt. By evapora-tion and calcination the foda will be left pure, and thus its quantity accu-rately determined.

If the water took up any thing dur-ing the boiling, in the third procefs, it muft be felenite. This the pure kali will precipitate.

The refiduum, on which neither the fpirit of wine nor the water could act, may confift of calcareous earth, aerated magnefia or iron, clay and quartz. The two laft are rare. A brown or yellow colour indicates iron; a white grey, the abfence of it. If it contain iron, it fhould be moif-tened and expofed to the rays of the

§ fun,

fun, and, when the iron is perfectly
rufted, digefted in diftilled vinegar.
This will diffolve the lime and magnefia, which may be feparated by the
vitriolic acid, as we have pointed out
above. The iron and clay are foluble in pure marine acid, from which
the former may be precipitated by the
Pruffian alkali; the latter, by the
mild vegetable alkali. The matter
which remains is ufually quartzofe:
this the blowpipe will afcertain.

DISTILLATION is employed to
procure the aeriform fluids contained
in water. For this purpofe fome
pounds muft be put into a retort, of
which they fhould not fill more than
half or two thirds: to the retort a
recurved tube is to be adapted, paff-
ing underneath an inverted veffel
filled with mercury. The retort is
then to be heated till the water boils,

or

or till no more elaftic fluid paffes over. Hepatic air, and fixed air, are thofe moft commonly met with in waters. The former is eafily diftinguifhable by its peculiar fmell; the latter by being abforbed by lime-water, from which it precipitates the calcareous earth.

AN
ACCOUNT
OF THE
MEDICINAL VIRTUES, &c.
OF
MINERAL WATERS.

ABCOURT, *near St. Germains, four leagues from Paris.*

IT is a brisk chalybeate water, impregnated with fixed air, and natron; and resembles the waters of *Spa* and *Ilmington*.

ABERBROTHOCK, *in Scotland.*

It is a chalybeate water, similar to those of *Peterhead* and *Glendy*.

ACTON,

ACTON, *near London, in the county of Middlesex.*

The wells are much frequented in May, June, and July.

The water is clear, and without smell, but its taste is somewhat bitterish.

It contains upwards of five drams of vitriolated magnesia in the gallon.

It is one of the strongest purging waters about London; and is noted for causing a great soreness in the fundament..

AGHALOO, or APHALOO, *in the county of Tyrone, Ireland.*

It is a sulphureous water of the same kind with that of *Swadlingbar*, but stronger. Like that, it is also impregnated with natron, and a small quantity of purging salt.

AIX-

Aix-la-Chapelle*, *in the duchy of Juliers, Germany.*

This place has long been famous for its hot fulphureous waters and

* My friend, the ingenious Dr. Simmons, F. R. S. who made many experiments on the waters during his refidence at this place, has favoured me with an account of their feveral temperatures, as repeatedly obferved by himfelf, with a thermometer conftructed by Nairne.

The fpring which fupplies the Emperor's bath (*Bain de l'Empereur*), the New Bath (*Bain Neuf*), and the Queen of Hungary's bath (*Bain de la Reine de Hongrie*) —	127°
St. Quirin's bath (*Bain de St. Quirin*)	112°
The Rofe bath (*Bain de la Rofe*), and the Poor's bath (*Bain des Pauvres*), both which are fupplied by the fame fpring —	112°
Charles's bath (*Bain de Charles*), and St. Corneille's bath (*Bain de St. Corneille*)	112°
The fpring ufed for drinking is in the High Street, oppofite to Charles's bath; the heat of it at the pump is — —	106°

Dr. Afh makes the greateft heat 136° of Fahrenheit, placing the temperatures of the different baths from 3° to 9° higher than in the above account.

baths.

baths. They arife from feveral fources, which fupply eight baths conftructed in different parts of the town.

Thefe waters near the fources are clear and pellucid, and have a ftrong fulphureous fmell refembling the wafhings of a foul gun; but they lofe this fmell by expofure to air. Their tafte is faline, bitter, and urinous. They do not contain iron. They are alfo neutral near the fountain, but afterwards are manifeftly, and pretty ftrongly alkaline, infomuch that cloaths are wafhed with them without foap.

The gallon contains about two fcruples of fea falt, the fame quantity of chalk, and a dram and half of natron.

They are at firft naufeous and harfh, but by habit become familiar and agreeable. At firft drinking alfo they generally affect the head.

Their general operation is by ftool
and

and urine, without griping or diminution of strength ; and they also promote perspiration.

The quantity to be drunk as an alterative, is to be varied according to the constitution, and other circumstances of the patient. In general, it is best to begin with a quarter, or half a pint in the morning, and increase the dose afterwards to pints, as may be found convenient. The water is best drunk at the fountain. When it is required to purge, it should be drunk in large and often repeated draughts.

In regard to bathing, this also must be determined by the age, sex, strength, &c. of the patient, and by the season. The degree of heat of the bath should likewise be considered. The tepid ones are in general the best, though there are some cases in which the hotter ones are most pro-

G   per.

per. But even in thefe it is beft to begin with the temperate baths, and increafe the heat gradually.

.. Thefe waters are efficacious in dif-eafes proceeding from indigeftion, and from foulnefs of the ftomach and bowels. In rheumatifms; in the fcurvy, fcrophula, and difeafes of the fkin; in hyfteric, and hypochon-driacal diforders; in nervous com-plaints and melancholy; in the ftone and gravel; in paralytic complaints; in thofe evils which follow an inju-dicious ufe of mercury, and in many other cafes.

They ought not however to be gi-ven in hectic cafes where there is heat and fever, in putrid diforders, or where the blood is diffolved, or the conftitution much broken down.

ALFORD,

ALFORD, or AWFORD, *in Somer-setfhire, about* 24 *miles fouthward of Bath.*

This falt fpring was difcovered in 1670, from the pigeons which flew thither in great numbers to drink the water: thofe birds being known to be fond of falt.

It contains a purging falt, together with a portion of fea falt.

It is ftrongly purgative.

It is recommended as cooling, cleanfing, and attenuating. As a good remedy in the fcurvy, jaundice, and other glandular obftructions. It alfo promotes urine and fweat, and therefore is good in gravelly and other diforders of the kidnies and bladder; and in complaints arifing from ob-ftructed perfpiration.

ALKER-

A L K E R T O N, *in Gloucestershire, near the city of Gloucester.*

It is a purging water, of the nature of those of *Dulwich* and *Epsom*.

A N A D U F F, *in the county of Leitrim, Ireland.*

It is a sulphureous water, of the same kind with those of *Killasher* and *Drumasnave,* but weaker.

A N T O N I A N.

See *Tonstein.*

A S H W O O D, *in the county of Fermanagh, Ireland.*

It is a sulphureous water; and contains natron, with a small quantity of purging salt.

In its virtues it resembles the waters of *Drumgoon* and *Swadlingbar.*

A S K E R O N,

ASKERON, *five miles from Don-caster, in Yorkshire.*

It is a strong sulphureous water, and is slightly impregnated with a purging salt.

A gallon contains forty-eight grains of vitriolated magnesia, with a little sea salt, and a dram and half of earth.

It is recommended internally and externally in strumous and other ul-cers, scabies, leprosy, and similar com-plaints.

It is good in chronic obstructions, and in cases of worms and foulness of the bowels.

It operates by stool and urine.

ASTROPE, *near Banbury, in Ox-fordshire.*

It is a brisk, spirituous, pleasant-tasted chalybeate water, and is also gently purgative.

It

It fhould be drunk from three to five quarts in the forenoon.

It is recommended as excellent in female obftructions, the gravel, hypochondria, and fimilar diforders.

A S W A R B Y, *feven miles from Grantham, in Lincolnfhire.*

It is a fine blueifh chalybeate water, and is gently laxative without occafioning griping or faintnefs, or a pain in the fundament; which is a common effect of waters impregnated with fea falt. In its virtues it refembles the *Cheltenham* water.

A T H L O N E, *in the county of Rofcommon, Ireland.*

It is a chalybeate water, without colour or fmell, but it will not keep.

It operates by urine, and is gently laxative. It feems to refemble the *Hartlepool* water.

A Y L E-

AYLESHAM, *in Norfolk.*

It is a flight chalybeate water, fimilar to that of *Iflington.*

BADEN, *in Auſtria, Germany.*

The waters are warm and fulphureous, and have been recommended in thofe diforders in which the *Bareges* and *Aix-la-Chapelle* waters have been found ferviceable. They are particularly fpoken of for the cure of gun-fhot wounds, and the complaints which remain after them.

BADEN BADEN, *in Swabia, Germany.*

There are a number of hot fulphureous fprings and baths in and near this place, which have the fame general virtues as thofe of *Aix-la-Chapelle* and *Bareges.* Taken inwardly they are alfo gently laxative.

G 4   BAG-

BAGNERES, *in the Bigorre, France.*

At this place are a variety of warm baths, which are ufed in the fame diforders as thofe of *Aix-la-Chapelle.*

The waters of fome fprings taken internally are diuretic, and others purgative.

BAGNIGGE WELLS PURGING WATER.

It is fituated on the north-eaft fide of London, near Iflington, and is much frequented in the fpring.

It is a falt purging water, containing in the gallon 257 grains of fea falt and vitriolated magnefia mixed.

Its virtues are fimilar to thofe of Pancras and Acton.

The dofe is from a pint to a quart. But it is ufually quickened with Glauber's, or other falts.

The

The CHALYBEATE WATER.

It is clear when it comes from the pump, and has a flight irony taste.

When first taken to the quantity of three or four glaffes, it is ufually purgative. But this effect does not continue after the inteftines are cleared of their vitiated contents.

In its virtues it refembles the *Orfton* and other fimilar chalybeates.

BALARUC, *in Languedoc, France.*

The waters of this place are hot, and gently purgative. They have been ufed in many diforders for which falt purging waters are prefcribed.

They contain calcareous and magnefian muriate, fea falt, and chalk.

As they are hot, they have alfo been found particularly ufeful in cafes where warm baths are proper, to affift the operation of fuch waters.

Hence

Hence they have been found particularly useful in palsies and rheumatisms, in scrophula, and many other disorders.

B A L E M O R E.

See Ilmington.

B A L L, or B A N D-W E L L, *in Lincolnshire.*

It resembles the *Dropping-Well* water. Four or five half pints are reckoned a sufficient dose.

B A L L Y C A S T L E, *in Antrim, Ireland.*

It is a chalybeate water, somewhat of the nature of those of *Islington* and *Hampstead*; only it is slightly sulphureous.

B A L L Y N A H I N C H, *in Down, Ireland.*

It is a very clear, cold, chalybeate
and

and fulphureous water, and is good in fcorbutic and cutaneous difeafes, in lofs of appetite, &c.

BALLYSPELLAN, *near Kilkenny, in Ireland.*

It is a flight chalybeate water, fimilar to thofe of *Iflington* and *Hampftead.*

BARÈGES, *in the Bigorre, France.*

There are feveral fprings of hot fulphureous water at this place, which form four baths *.

The water is at firft clear; but by ftanding throws up a thin pellicle, refembling a fine light oil. It has a flight fulphureous fmell, like that of eggs boiled hard. It has a foft and

* Dr. Simmons informs me, that on plunging his thermometer into the hotteft fpring the mercury rofe to 112°.

Dr. Afh placed the hotteft at 122°, the leaft hot at 97'.

fome-

somewhat naufeous tafte, and feels
foft, like foap-water, or oil. Its volatile parts fly off on expofure to the
air; and it is beft drunk at the fountain head.

It contains fulphurated kali, with
a very fmall portion of fea falt, natron, calcareous earth, and felenite.

This water operates by perfpiration,
and by urine; but feldom by ftool.
The dofe is ufually a quart, or three
pints.

It is alfo ufed as a bath; as a fomentation; and as a douche.

The Barèges waters have been recommended in a variety of diforders;
in rheumatifms, palfies, convulfions,
cutaneous eruptions, the gout, fcurvy,
&c. Alfo in wounds, ulcers, hard
tumours; and they are faid to have
been efficacious in old gun-fhot
wounds, and in hard knots in the ure-
thra after venereal complaints.

B A R-

BARNET and NORTH-HALL.

The *former* spring is situated at East Barnet in Hertfordshire.

The *latter* lies about three miles north of High Barnet.

They are both purging waters, somewhat of the nature of *Epsom* water, but much weaker. That of *Barnet* is the strongest of the two, containing five drams of vitriolated magnesia, with a little sea salt, in the gallon.

BARROWDALE. *The spring is about three miles from Keswick in Cumberland.*

It is a salt water, and much of the nature of that of the sea.

A gallon affords seven ounces and two drams of sea salt mixed with a little vitriolated magnesia.

It is a brisk and rough purge even

to

to ftrong conftitutions, occafioning great thirft, and heating the body. A pint is ufually fufficient for a dofe.

Taken in lefs quantity (half, or a quarter of a pint) it operates by urine.

It is of excellent ufe in fcorbutic complaints, in the King's evil, and the leprofy. It is alfo powerful in removing chronic obftructions; in clearing the blood of acrimonious-humours; in difeafes of the fkin; and in all thofe complaints in which fea water is ferviceable. Like that alfo it may be ufed externally by way of fomentation or bath. See *Sea Water*.

BATH, *in Somerfetfhire.*

This place has long been famous for its warm chalybeate waters. There are feveral fprings, but their waters are all of the fame nature. There are fix baths; but the principal are
the

the *King*'s bath, the *Queen*'s bath, and the *Crofs* bath. The others are only appendages to thefe. The two former raife the thermometer to 116°, the latter to 112°.

The water when viewed in the baths has a greenifh, or fea colour: but in a vial it appears quite tranf-parent and colourlefs, and it fparkles in the glafs.

It has a very flight faline, bitterifh, and chalybeate tafte, which is not dif-agreeable, and fometimes fomewhat of a fulphureous fmell; but this latter is not ufually perceivable, except when the baths are filling.

The gallon of Bath water contains twenty-three grains of chalk, the fame quantity of muriate of magnefia, thirty-eight of fea falt, and 8.1 of aerated iron.

As it rifes from the pump, it con-tains fixed air, or other volatile acid,

in

in a fufficient quantity to curdle milk and diffolve iron.

The Bath water operates powerfully by urine, and promotes perfpiration. If drank quickly, in large draughts, it fometimes purges; but if taken flowly and in fmall quantity, it rather has the contrary effect. An heavinefs of the head, and inclination to fleep, are often felt on firft drinking it.

This water when taken inwardly gives a brifk ftimulus to the nerves and fibres, and feems to give new life and vigour to the whole frame. It alfo powerfully corrects putrefcent acrimony. Hence when taken into the ftomach it is faid to dilute and blunt whatever putrefcent humours it meets with; while its brifk, volatile, chalybeate principles ftimulate and increafe the tone of the ftomach and bowels, and brace up their fibres and nerves.

nerves. Entering the circulation, they pervade the minuteft veffels; dilute, blunt, and correct thofe fluids in the blood which are too putrefcent; increafe the action of the whole vafcular fyftem to promote the circulation through the fmalleft veffels, to break down grofs humours, to remove obftructions, and to promote fecretions of the fkin and kidnies, for carrying off thofe fluids that are unfit to circulate longer in the general mafs. And hence it is that they have been found fo ferviceable in fuch a variety of diforders. In female complaints, for example; fuch as obftructions of the menfes; barrennefs proceeding from obftruction and relaxation of the womb; the fluor albus, &c. In hyfteric and hypochondriacal diforders; in complaints of the ftomach and bowels proceeding from weaknefs and laxity, or from

putref-

putrefcent humours. In pains of the ftomach, attended with bad digeftion, and in many cholicky and other dif-orders of the ftomach and bowels. In diforders of the head and nerves; fuch as palfies, epilepfies, convulfions, &c. In difeafes of the fkin; the fea fcurvy; leprofy. In obftructions of the liver, fpleen, and other bowels; in gouty and rheumatic complaints; in the ftone and gravel; and in many other difeafes.

Thefe waters being of an heating nature, it is ufual, previous to a courfe of them, to cool the body by gentle purges, by a low diet, and, if found neceffary, by bleeding.

They may be drunk from half a pint, to two, three, or four pints in a day, according to circumftances. The beft method is to take one, two, three, or four half glaffes at proper intervals in the morning; a glafs or two an hour
before

before dinner; and as much about the fame time before fupper. The patient in the mean time fhould live upon a light diet, eafy of digeftion; ufe proper exercife; go early to bed; and rife betimes in the morning.

In fome cafes, however, thefe waters are hurtful. In hectic fevers, for example; in fuppurations of the lungs; in fits of the gout; and in the rheumatifm if inflammatory; and indeed in all cafes of inflammation; as alfo where the action of the fibres is already too ftrong, the animal-heat too great, and the blood thick and fizy.

The quantity of the waters drank in a day fhould be gradually encreafed to as much as the patient can bear; and after continuing that quantity a fufficient time, it fhould be as regularly diminifhed. The courfe may

be

be continued for a month or fix
weeks.

The ufual. feafon for the Bath waters is in April, May, and June ; and
in Auguft, September, and October.

Thefe waters are alfo ufed externally in a variety of diforders, and
with good effect, either by bathing
or pumping, as occafion may. require; efpecially if ufed inwardly at
the fame time. Forefts of crutches
left there, are an ample teftimony of
the efficacy of bathing in paralytic
cafes. By foftening and relaxing the
parts, and at the fame time giving a
gentle ftimulus, they are alfo of fervice
in removing many inveterate gouty and
rheumatic complaints. In difeafes of
the limbs, &c. arifing from obftructions ; in fprained, relaxed, and ftiff
joints ; in fcorbutic and cutaneous difeafes, old fores and ulcers, and in
many other cafes ; and when the
                                    com-

complaint is local, pumping is generally preferred to bathing.

It is a certain effect of thefe and
other baths, to throw out a redness
and kind of eruption on the fkin, efpecially in thofe who are fcorbutic,
&c. But this effect difappears by
their continued ufe, and the diforders
themfelves are at length cured.

The mud and fcum of thefe waters
have alfo been applied with good effect by way of poultice in hard fwellings, in weak joints, in contractions
of the limbs, in fcald heads, running
ulcers, &c. and herbs are fometimes
boiled with them in the Bath water
to a proper confiftence, for thefe and
the like purpofes.

B I L T O N, *near Knarefborough, Yorkfhire.*

The water has a ftrong fulphureous
fmell,

fmell, and taftes fomewhat faltifh. It
is colder than common water.

It contains natron, with a little
fea falt.

It acts as a gentle purge; and is
fomewhat fimilar in virtue to the *Sutton Bog* water.

BINLEY, *near Coventry, Warwick-fhire.*

It is a chalybeate water, and alfo
purgative and diuretic. It refembles
the *Scarborough* water, but is lefs purgative.

BIRMINGHAM, *in Warwickfhire.*

Near this place is a brifk chalybeate water, which feems to refemble
that of *Hampftead* in *Middlefex.*

BORDSCHEIT, or BORSET*, *about a mile and half from Aix-la-Chapelle, Germany.*

The waters are warm, and of the nature of thofe of Aix-la-Chapelle, being, however, fomewhat more purgative; but they are only ufed as baths, for the difeafes in which the waters laft mentioned are recommended, and alfo in dropfical and oedematous cafes.

BRABACH, *in the diftrict of Mengerfkirchen, in the county of Naffau, Germany.*

It is a brifk fpirity chalybeate water, which may be preferved long in

---

* The waters at this place, which is only about a mile from Aix-la-Chapelle, are diftinguifhed into the upper and lower fprings. The former, which contain no hepatic air, were found by Dr. Simmons to raife the thermometer to 158°; the latter, all of which are fulphureous, to 127° only. All the baths are fupplied by the firft.

§   well-

well-ftopt bottles, though it foon fpoils in the open air. It has a fome-what falt, fulphureous, and aftringent tafte, and contains natron.

It refembles the German *Spa Water* in its general virtues.

BRANDOLA, *in Italy*.

It is a flight chalybeate water, ex-tremely limpid and cryftalline, im-pregnated with an alkaline falt, and abounding in fixed air. It fmells fomewhat fulphureous, and has an acidulous tafte.

It is commonly drunk from two pints to a gallon or more in a day. It promotes urine and perfpiration, and is gently laxative.

Its virtues feem to refemble thofe of the *Iflington* and *German Spa* wa-ters.

BRENT-

BRENTWOOD, *in Effex.*

It is a purging water, of the nature of thofe of Pancras, Epfom, and Dul-wich.

BRISTOL, *in Somerfetfhire.*

The fprings are known by the name of the *Hot Wells.*

The water at its origin is warm, clear, pellucid and fparkling; and if let ftand in a glafs, covers its infide with fmall air-bubbles. It has no fmell, and is foft and agreeable to the tafte. It raifes the thermometer from about feventy to eighty degrees. It contains $12\frac{3}{4}$ grains of chalk, $5\frac{1}{4}$ of muriate of magnefia, and $6\frac{1}{2}$ of fea falt in the gallon.

It has been recommended in a va-riety of diforders. In confumptions and weaknefs of the lungs; in cafes

H at-

attended with hectic fever and heat (in which, among other properties, it differs from the *Bath* water) in uterine and other internal hæmorrhages, and in immoderate difcharge of the menfes; in old diarrhœas and dyfenteries; in the fluor albus; in gleets; in the diabetes; and in other cafes where the fecretions are too much increafed, and the humours too thin; in the ftone and gravel; in the ftrangury; in colliquative fweats; in fcorbutic and fimilar cafes; in cholics; in the gout and rheumatifm; in lofs of appetite and indigeftion; and in many other difeafes.

The ufual method of drinking the water is a glafs or two before breakfaft, and about five in the afternoon. The next day three glaffes before breakfaft, and as many in the afternoon; and this is to be continued during the patient's ftay at the Wells.

A quar-

A quarter or half an hour is allowed between each glafs.

A courfe of thefe waters requires no preparation further than to empty the bowels by fome gentle purge; and **if** heat or fever require, to take away a few ounces of blood. Coftivenefs, however, fhould be avoided during the courfe.

Externally they are ufeful in fore and inflamed eyes; fcrophulous and cancerous ulcers; and in other fimilar cafes.

This water is cooling and quenches the thirft. It is beft drunk at the fpring head; though it will bear carriage tolerably well.

B R O M L E Y, *in Kent.*

It is a chalybeate water, refembling thofe of *Spa, Iflington,* and *Hampftead.*

BROUGHTON, *in the West Riding
of Yorkshire, near Coln, in Lancashire.*

It is a strong sulphureous water; it
turns silver and copper black; it red-
dens the leaves of trees, &c. and makes
the bottom of its bason black.

It is impregnated with sea salt, and
a purging salt; and its virtues are
similar to those of the *Harrowgate*
water.

BUCH, *situated about a German mile
from the Caroline baths in Bohemia.*

The waters have a brisk pungent
taste, and are plentifully impregnated
with *fixed air*. This, on exposure,
flies off, and they become insipid. In
this they differ from *Seltzer* water,
which acquires a lixivial taste by
standing.

They contain, however, natron, in
the

the proportion of about sixteen grains to the gallon; and therefore their virtues are similar to those of the *Tilbury* and *Seltzer* waters, but much weaker.

BUGLAWTON, *near Congleton, in Cheshire.*

It is a sulphureous water, impregnated with a purging salt, and in its virtues seems to resemble the *Askeron* water.

It is intensely cold, and has a pretty strong sulphureous smell and taste.

BURLINGTON, *in Yorkshire.*

It is a brisk chalybeate water, and resembles those of *Scarborough* and *Cheltenham*, tho' it seems to be less purgative.

BURNLEY, or BOURNLEY, *in Lancashire.*

It is a chalybeate water of the nature

ture of the *Scarborough*, but lefs purgative.

BUXTON, *in Derbyfhire.*

This is a hot water, refembling
that of *Briftol.* It raifes the thermometer to 81° or 82°.

It has a fweet and pleafant tafte.

It contains a little calcareous earth,
together with a fmall quantity of fea
falt, and an inconfiderable portion of
a purging falt. Iron has been difcovered in it, but in fo extremely fmall
a quantity as not to deferve notice:
and even that perhaps owing to accident.

This water taken inwardly is efteemed good in the diabetes; in
bloody urine; in the bilious cholic;
in lofs of appetite, and coldnefs of
the ftomach; in inward bleedings;
in atrophy; in contraction of the
veffels and limbs, efpecially from age;

in

in cramps and convulfions; in the dry afthma without a fever; and alfo in barrennefs.

Inwardly and outwardly it is faid to be good in rheumatic and fcorbutic complaints; in the gout; in inflammation of the liver and kidnies, and in confumptions of the lungs; alfo in old ftrains; in hard callous tumours; in withered and contracted limbs; in the itch, fcabs, nodes, chalky fwellings, ring-worms, and other fimilar complaints.

Befides the hot water, there is alfo a cold *chalybeate* water, with a rough irony tafte. It refembles the *Cawthorp* water.

CANNOCK, *near Stafford.*

It is one of the beft and lighteft chalybeate waters in Staffordfhire. In its virtues it refembles thofe of *Hampftead* and *Iflington.*

CAPE

CAPE CLEAR, *fituated in the moſt ſouthern part of Ireland.*

It is a ſmooth, ſaltiſh water, and lathers with ſoap.

It contains about half a dram of natron, mixed with a little ſea ſalt, in the gallon.

Its virtues are ſimilar to thoſe of the waters of *Tilbury* and *Clifton*, but weaker.

CARGYRLE, *in Wales.*

The ſpring is ſituated about ten or twelve miles from Cheſter.

The water is of the nature of the *Barrowdale* water, but much weaker, ſeveral quarts being required to be taken for a purge.

CARLTON, *near Newark upon Trent, in the county of Nottingham.*

It is chalybeate water, reſembling thoſe of *Iſlington* and *Hampſtead*, but

it

it has a fœtid fmell, like infufion of
horfe-dung.

CAROLINE BATHS, *at Carlfbad,*
in Bohemia, Germany.

The waters of this place are hot.

They contain thirty-fix grains of
chalk, forty-eight grains of fea falt,
one hundred and two grains of natron,
and fix drams of vitriolated natron.
They are alfo impregnated with iron.

The higheft temperature is 165°,
the loweft 114°.

They are recommended externally
and internally in female obftructions;
in relaxed habits; in glandulous obftructions; in diforders arifing from
vifcid fluids, and in a variety of other
complaints; and it is faid, that they
may be drunk, and bathed in, by perfons of all ages and conftitutions, with
fafety.

CARRICKFERGUS, *in the county of Antrim, Ireland.*

The water is of a blueish colour, and a very foft tafte, at the fountain-head.

It is weakly purgative; and muft be drank to the quantity of two or three quarts.

Near this fpring is another, a gallon of the water of which affords about an ounce and half of fea falt, and a little vitriolated magnefia, with a quantity of an earthy matter.

CARRICKMORE, *in Ireland.*

It is fituated about five miles from Belturbet, in the county of Cavan.

The water has a foft, milky tafte, like Briftol water; and putrifies by keeping.

It curdles a folution of foap; and with falt of tartar gives a white fediment.

§   It

It contains natron, together with a purging falt.

Its virtues therefore are fimilar to thofe of *Tilbury* and *Clifton.*

CASHMORE, *in the county of Waterford, Ireland.*

It is near the *Crofs-town* water, which it refembles in virtues, though ftronger.

CASTLECONNEL, *in the county of Limerick, Ireland.*

It is a chalybeate water of confiderable repute, and refembles the German *Spa* waters.

CASTLE LOED, *in Rofsfhire, Scotland.*

This is a ftrong fulphureous water. The gallon yielded near $1\frac{4}{5}$ grains of abforbent earth, $26\frac{3}{5}$ of felenite, $30\frac{3}{5}$ of faline matter confifting of vitriolated

H 6   natron

natron with a little fulphur, and pro-bably a fmall portion of marine bittern.

It has been many years in repute againft cutaneous difeafes.

CASTLEMAIGN, *in the county of Kerry, Ireland.*

It is a fulphureous, and ftrongly chalybeate water, and in its virtues feems to refemble that of *Deddington.*

CAWLEY, *near Dranefield, in Der-byfhire.*

It is fulphureous, and gently pur-gative; and refembles the *Afkeron* water.

It contains about half a dram of vitriolated magnefia in the gallon.

CAWTHORP, *four miles from Bourne, in Lincolnfhire.*

It is a faltifh chalybeate water, and foams much as it rifes from the fpring.

It

It refembles the *Tunbridge* water in virtues, but is faid to be more purgative; and is alfo a good corrector of acidities.

CHADLINGTON, *near Chipping-Norton, Oxfordfhire.*

The water has a faltifh tafte, and fmells like the wafhings of a foul gun.

It is one of the waters termed fulphureous.

It contains alfo natron, together with a little fea falt.

It acts as a purgative; and its virtues refemble thofe of the *Sutton Bog* water.

CHAUDE FONTAINE, *about two leagues from Liege, and three from Spa, in Germany.*

The water of thefe fprings is hot, and fupplies fifty baths.

It is claffed by authors with the fulphureous waters; but Dr. Sim-

5   mons,

mons *, who fpent fome time at this place in 1776 and 1777, informs me they have no fulphureous fmell; that they are impregnated with calcareous earth, and natron, and alfo with fixed air.

They are not chalybeate; and therefore rather refemble our *Briftol* and *Buxton* than the *Bath* water.

Their virtues *externally* however may be collected from what has been faid of the *Aix-la-Chapelle* and *Bath* waters.

CHELTENHAM, *in Gloucefterfhire, fix miles from Gloucefter.*

It is one of the beft and moft noted purging chalybeate waters in England, though it is not fo much frequented as formerly.

* The fame gentleman informs me, that on the 5th of July 1777, when the mercury in his thermometer, in the fhade, ftood at 75°, it rofe in the bath to 92°.

The

The gallon contains eight drams of a purging falt, partly vitriolated natron, partly vitriolated magnefia; twenty-five grains of magnefia, part of which is united with marine, part with aerial acid; and nearly five grains of iron combined with aerial acid. It alfo yielded thirty-two ounce mea-fures of air, twenty-four of which were fixed air, the reft phlogifticated with a portion of hepatic air.

The dofe is from one pint to three or four. It operates with great eafe, and is never attended with gripings, tenefmus, or ftraining at ftool. It is beft taken a little warm.

It alfo creates an appetite; is excel-lent in fcorbutic complaints, and has been ufed with fuccefs in the gravel.

As the fpring has been calculated to yield only thirty-five pints of water an hour, without frugal management there would not be enough to fupply

the

the demands of the drinkers. The Walton water has lately been recommended as a fubftitute to obviate this inconvenience.

CHIPPENHAM, *in Wiltfhire*.

It is a pretty ftrong chalybeate water, refembling thofe of *Iflington* and *Tunbridge*.

CLEVES, *in the duchy of Cleves, Germany*.

It is a brifk chalybeate water, and operates by urine. It refembles the *Pyrmont* water.

CLIFTON. *This is a village near Deddington, in Oxfordfhire*.

The well is about a furlong fouth of Clifton. The water is clear, and has but little tafte.

The principal ingredient in it is natron, of which about fixty-five grains are contained in the gallon.

Its

Its virtues are fimilar to thofe of the *Tilbury* water, though in a lefs degree. But as it alfo contains a purging falt, it is more purgative than that.

It has been much ufed by way of bath in diforders of the fkin.

COBHAM, *in the county of Surry.*

It is a chalybeate water, of the nature of that of *Tunbridge*, but rather ftronger of the iron.

There is alfo a purging water near it, from a gallon of which Dr. Hals obtained an ounce or upwards of a refiduum, confifting principally of vitriolated magnefia.

CODSALWOOD, *five miles from Wolverhampton, Staffordfhire.*

It is a ftrong fulphureous water.

In its virtues it feems to refemble the *Afkeron* water.

COL-

COLCHESTER, *in the county of Essex.*

It is a purging water of the nature of thofe of *Acton* and *Epfom.*

COLURIAN, *in the parifh of Ludg-van, in Cornwall.*

It is a chalybeate water, and feems to refemble thofe of *Hampflead* and *Iflington.*

COMNER, or CUMNER, *in Berkfhire, four miles weft of Oxford.*

The water is of a whitifh colour, efpecially in the fummer.

It contains two hundred and forty-four grains of vitriolated magnefia with excefs of the magnefia, and fifty-two of chalk.

It is purgative, and may be drunk to the quantity of one, two, or three quarts,

quarts, according to the patient's conftitution.

COOLAURAN, *in the county of Fermanagh, Ireland.*

It is a chalybeate water, refembling
that of *Peterhead*, but weaker.

CORSTORPHIN, *two miles from Edinburgh, Scotland.*

It is a weak fulphureous water,
very flightly impregnated with fea falt,
and vitriolated magnefia.

There is another fpring, about a
mile from Edinburgh, on the banks
of the water of Leith.

They refemble the *Moffat* water in
virtues ; and the latter is reckoned the
ftrongeft.

COVENTRY, *in Warwickfhire.*

It is a chalybeate and purging water,
which fits eafy upon the ftomach,
foon

foon paffes off, raifes the fpirits, and creates an appetite.

In its general virtues it refembles the *Scarborough* and *Cheltenham* waters.

CRICKLE SPA, *fituated near Broughton, in Lancafhire.*

It is a ftrong fulphureous water, a a gallon of which contains about four drams and half of fea falt and vitriolated magnefia, the former of which is greatly predominant, and about fifty grains of calcareous earth.

It is purgative; and in its virtues refembles the *Harrogate* water.

About a mile diftant is Broughton water, of the fame nature, but containing lefs fea falt.

CROFT, *in the North Riding of Yorkfhire, on the confines of Durham.*

This is a ftrong fulphureous water, a gallon of which contains one hundred

dred

dred and fifty grains of calcareous earth, thirty of vitriolated magnesia, and ten of sea salt.

It is clear and sparkling, and its stream does not rise or fall by rain or drought.

It is purgative, and of the nature of the *Afkeron* water; and is said to have performed remarkable cures.

CROSS-TOWN, *near the town of Waterford, Ireland.*

It resembles the *Hartfell* water in Scotland.

This water vomits some, purges others, and with others operates by urine.

It seems at some times to contain a greater quantity of acid than at others.

CUNLEY-HOUSE, *near Whaley, in Lancashire.*

It is strongly sulphureous, and gently

ly purgative, and feems to refemble in its virtues the *Aſkeron* water.

D as W i l d - b a d, *within the walls of the town of Nuremberg, Germany.*

It is a chalybeate water, with a fub-aftringent tafte, and contains alfo a faline matter.

It has been recommended in ob-ftructions of the vifcera, and in female complaints.

D' A x en F o i x, *about fifteen leagues weſt of Thouloufe, France.*

This place abounds with hot ful-phureous waters of different tempera-tures. They are recommended as baths, or otherwife, in thofe com-plaints in which the *Aix-la-Chapelle* and *Barèges* waters are ferviceable.

D e d d i n g t o n.

This is a fulphureous chalybeate water;

water; but foon lofes its fulphureous
fmell by keeping.

Drunk in large quantities it is purgative; and in lefs dofes as an alterative, it is good in fcorbutic and cutaneous diforders.

DERBY, *near to the town of Derby, in Derbyfhire.*

It is a chalybeate water of the nature of that of *Tunbridge*, but feems
to be ftronger.

DERRINDAFF, *in the county of Cavan, Ireland.*

This is a fulphureous water, impregnated with a purging falt.

Its virtues refemble thofe of the
Afkeron water.

DERRYHENCE, or DERRYINCH, *in the county of Fermanagh, Ireland.*

The water is fulphureous.

It

It alfo contains natron, and refembles in its virtues the waters of *Drumgoon* and *Swadlingbar*.

DERRYLESTER, *in the county of Cavan, about three miles from Swadlingbar, Ireland.*

The water is of the nature of that of Drumgoon, but contains much lefs of the falts.

DOG AND DUCK.

A noted tea-drinking houfe in St. George's Fields, near London; and in the fpring and fummer months the waters are much reforted to.

The water is clear, and has but little tafte.

It is a mild purgative, containing vitriolated magnefia mixed with fea falt, and may be drank to the quantity of feveral pints. Moft frequently, however, it is quickened by the
addition

addition of Glauber's, or other purging salts.

It is of use in scrophulous complaints, leprosies and diseases of the skin; and is also said to prevent the return of cancerous diseases. For these complaints it may be used both internally and externally.

It is cooling and diuretic; and may be given freely to young people of robust constitutions. But it cools and relaxes people in years and of weak habits too much. It is also apt to bring on or increase the fluor albus in weakly women.

DORTSHILL, *near Litchfield, in Staffordshire.*

The water is a brisk chalybeate, similar to that of *Tunbridge.*

There is also a saline purging water of the nature of the *Barrowdale* water, but much weaker.

I   DRI-

DR I·BU R G, *about half a mile from the town of Driburg, in Weftphalia.*

The water, which is in the higheft reputation abroad, very much refem-bles the *Pyrmont*; containing the fame ingredients, but in a rather lar-ger proportion.

The quantity of fixed air obtained from it by Dr. Higgins was to that of Pyrmont as thirteen to twelve.

DR I G W E L L, *near Revenglas, in Cumberland.*

This is a brifk, fpirituous, fulphu-reous chalybeate; and in its virtues refembles the *Deddington* water.

DR O P P I N G W E L L, *at Knaref-borough, in Yorkfhire.*

It is very cold, limpid, and fweet-tafted; and in time petrifies fubftan-ces thrown into it.

In

In its virtues it refembles the *New-ton Dale* water. The dofe has for-merly been feveral quarts in a day; but three or four half pints are now judged fufficient.

Its ufe fhould be preceded by a dofe or two of rhubarb.

DRUMASNAVE, *called likewife Mount Campbell, in the County of Leitrim, Ireland.*

This is one of the ftrongeft ful-phureous waters in Ireland, as is fhewn by its quick and ftrong effect in difcolouring metals.

It is perfectly clear and limpid in common; but before rain becomes white.

It contains about twelve grains of natron, with a fmall quantity of purg-ing falt, in the gallon.

It operates powerfully by urine,

I 2 and

and purges fome conftitutions, but is faid to render others coftive.

DRUMGOON, *in the county of Fermanagh, Ireland.*

The water has a ftrong fulphureous fmell, and tinges filver of a copper colour in a few minutes. It alfo depofits a black fediment at the bottom of the well.

It contains near a dram of natron in the gallon, with a little fea falt.

It is recommended for the cure of cutaneous and fcrophulous diforders; and for worms.

There are two other fulphureous fprings in the neighbourhood; the one nearly refembles this; the other is more of a purgative nature.

DUB

DUBLIN SALT SPRINGS. *There are five of these Springs in* Francis Street, *and one in* Thomas's Court.

The waters are salt, and of the nature of *Barrowdale* water. For a purge, they must be taken to the quantity of several pints. They operate without griping.

DULWICH. *The Spring is situated between Dulwich and Lewisham, in the county of Kent.*

The water is clear, and has a brackish taste, leaving a bitterness in the throat.

It contains a purging salt, together with sea salt.

This is a celebrated purging water; is also diuretic; and is recommended in a variety of disorders.

It is particularly of use in complaints arising from obstructions; as

I 3 those

thofe of the liver, fpleen, and other vifcera.

It is recommended in the green ficknefs, the jaundice, the fcurvy, in difficulty of urine, and in gravelly complaints.

It is faid to ftrengthen the ftomach, and create a good digeftion.

It is alfo faid to ftrengthen the nervous fyftem, and therefore to be ferviceable in palfies, apoplexies, and other nervous diforders. In thefe cafes it is beft taken warm.

The courfe of drinking this water is ufually twenty days. Three pints a day are to be drunk at firft; it fhould be increafed to eight pints by the tenth day, and afterwards decreafed in the fame manner.

A new fpring has fince been difcovered, whofe virtues are fimilar to thofe of the old one, but it is ftronger.

Dun-

DUNNARD, *about eighteen miles from Dublin.*

This is a chalybeate water, refembling that of *Peterhead*, but weaker.

DUNSE, *in Scotland.*

It is a chalybeate water, fimilar to that of *Tunbridge.*

DURHAM. *The Spring is fituated near Durham, on the north fide of the river Ware.*

It is a ftrong fulphureous water, and is alfo impregnated with fea falt, of which it contains thirty-eight grains in the gallon.

In its virtues it refembles the *Harrogate* water.

Near to this, in the middle of the river, is a falt fpring, which is drunk as a purging water.

EGRA, *in Bohemia.*

This is a fpirity chalybeate water,

and

and operates both by ftool and urine. It contains lefs fixed air than the *Pyr-mont* water, but is more purgative. It abounds with vitriolated magnefia mixed with muriate of magnefia.

ENGHIEN, or ANGUIEN, *a city of Hainault*.

This water contains fulphur, vi-triolated magnefia, chalk, and mag-nefia.

EPSOM, *in Surry, about fixteen miles from London*.

The water has a flight faline tafte, is clear, and without fmell. But if it be kept in covered veffels for fome weeks in the fummer it will ftink, and acquire a naufeous and faltifh bitter tafte.

This was the firft water from which the bitter purging falt (thence called *Epfom falt)* was obtained. But the

falt

falt ufually fold by that name is different from that yielded by the Epfom water, though perhaps not inferior in virtue. It is made from the
bittern left after the chryftallization
of common falt from fea water.

The Epfom water is purgative;
for which purpofe it muft be drank to
the quantity of two or three pints. It
alfo operates by urine.

Taken in lefs quantity (about the
third part of a pint three times a day)
it is a mild alterative, and good in
thofe complaints for which the *Acton*
and *Pancras* waters are recommended.

It is likewife efteemed good for
wafhing old fores.

FAIRBURN, *in the county of Rofs,*
in Scotland.

This is about two miles from the

I 5   Caftle-

Castle-Loed well, which it nearly resembles, but is somewhat weaker.

FELSTEAD, *in Effex.*

The spring is situated at the bottom
of a rock.   The water is a light chalybeate, resembling that of *Islington*.

FILAH, *near Scarborough, in York-shire.*

This is a salt chalybeate water, and
is used by the common people as a
purgative; for which purpose they
drink to the quantity of several quarts:
it also operates by urine.

FRANKFORT, *in Germany.*

There are two strong sulphureous
waters in the neighbourhood of Frank-
fort on the Maine.

The one is called FAULPUMP,
                                    The

The other Fons Scabiosorum.

They are alfo impregnated with fea falt, and are of the nature of the *Moffat* and *Harrogate* waters.

GAINSBOROUGH, *in Lincolnfhire.*

This is a weak fulphureous chalybeate water, fimilar to that of *Deddington.*

GALWAY, *in the county of Galway, Ireland.*

It is a chalybeate water, of the nature of that of *Tunbridge.*

GLANMILE, *near Naul, in Ireland.*

It is a chalybeate water, refembling that of *Peterhead,* but weaker.

GLASTONBURY, *in Somerfetfhire.*

This water is of the fame nature

with thofe of *Tilbury* and *Clifton*; but weaker than either of thefe.

It has alfo a fmall mixture of fea falt.

It is naturally fweet, but by keeping becomes putrid.

This water was formerly in great repute; and many fuperftitions were held concerning it; but it has not lately been efteemed.

GLENDY, *in the county of Mairns, Scotland.*

This is a ftrong chalybeate water, little inferior to that of *Peterhead.*

GRANSHAW, *near Dunnaghadee, in the county of Down, Ireland.*

It is a chalybeate water, of the nature of that of *Caftle Connel.*

GROS-

GROSSENENDORF, *about five leagues from Hanover.*

This is a cold fulphureous water, of fome repute in the gout, palfy, and difeafes of the fkin and breaft.

The Landgrave of Heffe Caffel, in whofe dominions it is, has lately directed baths and other conveniences to be built here for the accommodation of invalids.

GUGGA.

See *Kuka.*

HAIGH, *near Wigan, in Lancafhire.*

It is impregnated with green vitriol; and is of the nature of the *Shadwell* water; which fee.

It works plentifully by vomit, and ftool; and is excellent for ftopping inward bleeding.

HAMP-

HAMPSTEAD.

This is a chalybeate water, of the nature of that of *Iſlington*, but ſomewhat ſtronger. The doſe is from half a pint to ſeveral pints.

It was formerly, and perhaps deſervedly, in great repute.

This water is better in the morning than in the middle of the day; and in cold weather it is much ſtronger than in hot.

HANBRIDGE, *in Lancaſhire.*

It is a chalybeate water, of the nature of that of *Scarborough*, but leſs purgative.

HANLYS, *near Shrewſbury, in Shropſhire.*

The water is clear, and limpid, and has a ſaline and bitter, though not diſagreeable taſte.

It ſprings up with impetuoſity at the

the fountain; and does not change colour, or lofe its virtue, by being expofed to the air.

It is purgative; and the dofe is from two to four half pints.

The gallon yields one hundred and twenty grains of vitriolated magnefia.

At this place there is alfo a *chalybeate water*. It is near to the purging water, and is of the nature of thofe of *Scarborough* and *Landrindod*.

It is brifk and pungent to the tafte, and as it is taken from the fountain, clear, and not unpleafant; but lofes its virtues by keeping.

H A R R O G A T E, *near Knarefborough, Yorkfhire.*

There are four fprings at this place, but the waters of all of them are nearly alike, except in the quantity of the faline matter they contain.

Of

Of the three old fprings, the higheft gave three ounces of folid matter; the loweft, an ounce and half; and the middle one, only half an ounce. Of the latter one hundred and forty grains were earth.

The water as it fprings up is clear and fparkling, and throws up a quantity of air-bubbles.

It has a ftrong fmell of fulphur, and is fuppofed to be the ftrongeft fulphureous water in England.

It has a falt tafte, as it contains a confiderable quantity of fea falt, together with a little marine falt of magnefia, and calcareous earth.

It is purgative; and the dofe required for this purpofe is about three or four pints.

When drank in fmaller quantities, it is a good alterative, and is found ferviceable in the fcurvy, king's evil, and difeafes of the fkin. It may be

ufed

used at the same time outwardly, by way of bath, or fomentation.

It has been found efficacious in deftroying worms.

It has been recommended in the gout, jaundice, the spleen, the green sickness, and other diforders arifing from obstructions.

It is used externally for removing old aches, strains, paralytic weaknesses, and the like. Also for the cure of ulcers, scabs, the itch, &c.

N. B. Between *Harrogate* and *Knaref-borough*, are also feveral *chalyb-eate* waters, which feem to refem-ble thofe of *Hampstead* and *Islington*. The moft remarkable are, *the Allum Well, the Sweet Spa,* and *the Tuewhet Well.* The latter is the ftrongeft.

HART-

HARTFELL, *in the county of An-nandale, Scotland.*

It is impregnated with green vi-triol, and refembles the *Shadwell* wa-ter, but is much weaker.

It is recommended in inward bleed-ings, in immoderate flux of the men-fes, in dyfenteries, in bloody urine, in the fluor albus, in gleets, in com-plaints of the ftomach and bowels, and in confumptions.

The dofe is from a gill to a pint or two, taken at repeated draughts in the morning.

Externally, it cures itchy, and tet-terous eruptions, and old fores, efpe-cially if taken at the fame time as an internal remedy.

HARTLEPOOL, *in Durham.*

This is a fine clear chalybeate wa-ter;

ter; with a flight fulphureous fmell, and pleafant tafte.

It is alfo diuretic and laxative; and is recommended as excellent in fcorbutic complaints, in bilious and nervous cholics, in pains of the ftomach and indigeftion, in the gravel, in female obftructions, in the hypochondriacal difeafe, in cachexy, in hectical heats, and in recent ulcers.

Holt, *near Bradford, in Wiltfhire.*

The water is limpid, and has but little tafte.

It contains a purging falt, together with a large quantity of earth.

On account of the latter ingredient, it is but a very mild purge, and two quarts are ufually required to produce any confiderable operation this way.

Taken in lefs quantity it is alterative, and diuretic.

It is alfo good as a diluent, cooler, and

and ſtrengthener; and creates an appetite.

Externally, rags, or ſpunge dipt in it, are ſaid to cure ſcrophulous ulcers, attended with carious bones; an internal courſe being obſerved at the ſame time.

It is alſo of ſervice in old running ulcers of the legs, and other parts; in cutaneous foulneſſes, tho' attended with hot corroſive humours; in the piles, in cancerous ulcers, and in foreneſſes of the eyes. But in theſe caſes alſo it muſt be uſed both internally and externally.

HOLT, *near Market-Harborough, in Leiceſterſhire.*

See *Nevil-Holt.*

JESSOP'S WELL, *on Stoke Common, near Cobham, in Surry.*

This is a ſtrong purging water,

with

with a naufeous tafte, and is alfo flightly chalybeate.

Drank to about a quart, it purges brifkly without griping, and operates likewife by urine.

It alfo enlivens the fpirits, and as the dofe is fmaller than that of other purging waters, it fits better on the ftomach.

It lofes its virtues by being kept.

Taken in lefs dofes as an alterative, it is a good antifcorbutic.

ILMINGTON, *in Warwickſhire, on the borders of Worceſterſhire.*

This a very clear and fparkling chalybeate water, abounding in fixed air, and impregnated alfo with natron.

It preferves its virtues for feveral weeks in bottles well corked; though if expofed to the air, it lofes them in twenty-four hours.

It

It operates by urine; and it alſo ſometimes purges.

It is recommended as excellent in ſcorbutic complaints, in obſtructions of the liver and ſpleen, in the jaundice, in beginning dropſies, in the gravel, and obſtruction of urine, and in diſorders ariſing from acidity.

Externally, it is good for old running ſores, ſcorbutic eruptions, and the like.

INGLEWHITE, *in Lancaſhire.*

It is a ſtrong chalybeate ſulphureous water, and is good in ſcorbutic, and cutaneous diſeaſes. But it will not purge unleſs Glauber's, or ſome other ſalt be added to it.

ISLINGTON, *in the county of Middleſex, near London.*

This is a ſlight chalybeate water, ſtriking

ftriking a purple or blackifh colour with galls, and is reckoned one of the beft of the kind about London.

The iron in this water is held in folution by means of *fixed air*, or *aërial acid*, as in the Pyrmont water. If, after the fixed air has efcaped, and the iron (which it held in folution) precipitates, the water be left to putrify, the fixed air difengaged by the putrefaction again diffolves the iron, and caufes it to be fufpended in the water; it then recovers its chalybeate tafte, and property of tinging black with galls, both which it had loft before.

It is recommended in indigeftion, and lofs of appetite, in lownefs of fpirits, nervous, hyfteric, and hypochondriacal complaints, and relaxed conftitutions, and raifes the fpirits greatly. It is good in the fluor albus, in weakneffes from mifcarriage, in obftructions of the liver, the kidnies,
&c.

4

&c. It is alfo ferviceable in difeafes of the fkin, in fcorbutic complaints, in the gravel, and in paralytic difor- ders.

It operates chiefly by urine, and may be drunk to the quantity of fe- veral half pints, or even pints, ac- cording to the patient's conftitution.

This water was formerly in great repute, and deferves to be more fre- quented than it is at prefent.

KANTURK, *in the county of Cork, Ireland.*

It is a chalybeate water, of the na- ture of that of *Peterhead*, but weaker.

KEDDLESTONE, *in Derbyfhire.*

This is a ftrong fulphureous wa- ter, and ftinks intolerably.

It is extremely clear at the fountain, but by ftanding becomes blackifh. It presently prefently turns filver of a black copper colour.

It contains thirty-eight grains of fea falt, and forty-two grains of calcareous earth in a gallon.

Its virtues refemble thofe of the *Harrogate* water.

KENSINGTON, *in the county of Middlefex, near London.*

It is a purging water, of the nature of thofe of *Acton* and *Pancras*.

KILBREW, *in the county of Meath, Ireland.*

This is a ftrong vitriolic chalybeate water, and refembles the *Shadwell* water.

Half a pint vomits and purges.

When taken as an alterative it fhould be ufed with great caution, beginning with a fmall quantity, and increafing the dofe.

K   It

It is recommended in the fluor albus, in immoderate fluxes from the womb, in obftinate intermittents, and in dropfies.

KILBURN, *in Middlefex, near London.*

It is a purging water, like thofe of *Bagnigge Wells, Dulwich,* &c.

KILLINGSHANVALLY, *in the county of Fermanagh, Ireland.*

This is a chalybeate water, and is alfo diuretic and gently laxative. It feems to refemble the *Hanlys* chalybeate water.

KILLASHER, *in the county of Fermanagh, Ireland.*

The water is ftrongly fulphureous, and contains natron.

Its virtues refemble thofe of the *Swadlingbar* water.

KILROOT, *in the county of Antrim,*
Ireland.

It is of the nature of *Barrowdale*
water, but weaker; feveral pints being
required for a purge.

KINALTON, or KYNOLTON, *a*
village in Nottinghamſhire.

The water is limpid and cooling,
with a fomewhat faltiſh tafte.

It is purging; but is weaker than
the Epfom water, and therefore muſt
be drunk plentifully.

A gallon contains about one hundred and fifty grains only of vitriolated magnefia.

KINCARDINE, *in the county of*
Mairns, Scotland.

This is a chalybeate water, little
inferior in ftrength to that of *Peter-*
head.

KINGSCLIFF, *in Northamptonſhire*.

It is a chalybeate laxative water, and reſembles the *Scarborough* and *Cheltenham* waters.

KIRBY, or KIRKBY-THOWER,
　　　　in Weſtmoreland.

There are two ſprings nearly of the ſame kind, only the lower one is reckoned the ſtrongeſt chalybeate.

The water of both is clear, fine, and has a chalybeate ſweetiſh taſte. Drunk to the quantity of ſeveral quarts it is purgative. It is alſo a good corrector of acidities.

KNARESBOROUGH,
See *Dropping Well.*

KNOWSLEY, *in Lancaſhire.*

This is a ſlight acidulous chalybeate water, and both taſtes and ſmells of iron.

If

If drunk to four or five pints it is laxative.

It refembles the *Scarborough* and *Cheltenham* waters in virtue, though it feems to be lefs purgative.

KORYTNA, *near Hunnobroda, in Moravia, Germany.*

It is fituated on a high and almoft inacceffible rock, in the midft of a thick wood.

It has a very fœtid difagreeable tafte, and a black colour; and there is much mud at the bottom of the well.

It is reckoned the ftrongeft fulphureous water in that country.

KUKA, *in the county of Graditz, in Bohemia, near the town of Jaromitz, at the conflux of the rivers Elbe and Orlitz, Germany.*

This is a very brifk chalybeate wa-

K 3   ter,

ter, highly impregnated with fixed air, and alfo with natron. It has a grateful and fomewhat pleafant tafte, and a pungent fmell, which affects the whole head. If it be heated, it emits a penetrating acid fulphureous fmelling vapour. It will not bear carriage.

It operates chiefly by infenfible per-fpiration; and fometimes by fpitting, by fweat, and by urine.

In its general virtues it refembles the *German Spa* waters.

LA MARQUISE, et LA MARIE.
See *Vahls*.

LANCASTER, or SALE's SPA,
in Lancafhire.

· This is a clear chalybeate water, powerfully diuretic, gently purgative, and vomits if taken to the quantity of feveral quarts.

Taken

Taken as an alterative it has the general virtues of the *Tunbridge* water.

LATHAM, *in Lancaſhire.*

It is a fine chryſtalline chalybeate, of the nature of the *Tunbridge* water.

LLANDRINDOD, *in the county of Radnor, South Wales.*

In this place there are three mineral ſprings :

1ſt. *The ſaline pump*, or *purging water.*

2d. *The ſulphureous water*, commonly called *the black ſtinking well.*

3. *The chalybeate rock water.*

The ſaline purging, or *pump water*, may be uſed as a *purge* twice in a week. It is directed to be drunk at the fountain-head by half pints, till it begins to operate; the patient walking or riding about between each draught. It operates alſo by urine.

For

For an *alterative*, about three pints
are directed to be drunk in a day. A
pint and half in the morning before
breakfaft, at three draughts, a quarter of an hour between each half
pint. The other pint and half likewife at three draughts : one an hour
before dinner; another about fix
o'clock in the evening; and the third
going to bed. If the body remain
coftive, the quantity muft be increafed. The courfe fhould be continued
feveral weeks; and the moft proper
feafon is the fummer.

It is alfo ufed as a bath and fomentation.

It is recommended both internally
and externally in the fcurvy, leprofy,
tetters, King's evil, and all cutaneous
foulneffes.

It is alfo prefcribed in the gravel,
the hypochondriacal difeafe, indigeftion, and in other complaints.

The

The sulphureous water; called also *the black stinking water,* from its strong smell, and the blackness of the channel through which it passes.

The quantity to be drunk cannot in general be determined: but it is best to begin with small doses, from a pint to a quart in the morning, taken at repeated draughts. The quantity may be increased as the constitution will bear; or as much as will sit easy on the stomach, and pass off well. When it gives the least uneasiness, it is a sign that the dose is too large.

It is also used outwardly, by way of bath or fomentation.

It is recommended in a variety of complaints. In the King's evil, scurvy, leprosy, and all cutaneous diseases; in the jaundice, hypochondriacal, and other disorders arising from obstruction; in the gravel, rheumatism, gout, bloody flux, hectic

K 5 fever,

fever, weakneffes of the limbs, want of digeftion, and many others.

The chalybeate, or *rock water,* is limpid and tranfparent, as taken from the fountain, but on ftanding foon lofes thefe qualities, together with its cha‑lybeate tafte. Mixed with fugar and rough cyder as it is taken up from the fpring, it excites a brifk fermentation.

It is recommended in fuch chronic diftempers as proceed from laxity of the fibres, and weaknefs of the mufcular fyftem; in weaknefs of the nerves; in paralytic complaints; and the like.

It is alfo good in fcorbutic cafes; in moift and convulfive afthmas; in obftinate agues; in obftructions of the lower belly; in wandering, flow, nervous fevers; and in diforders arifing from obftruction.

L L A N-

LLANGYBI, *in Caernarvonſhire,*
North Wales.

The water has a harſh taſte, in-
clining to bitter.

It has been found efficacious in
diſorders of the eyes; in the King's
evil; ſcald heads; ulcers; eruptions
of the ſkin; the ſcurvy; the itch, &c.
Alſo in rheumatiſms, palſy, and con-
vulſion fits.

This water has long been in repute
in the neighbourhood.

LEAMINGTON.

This is of the nature of *Barrow-*
dale water, but much weaker, containing little more than a fourth of the
ſame ingredients in an equal quantity.
The doſe for a purge is from a quart
to four or five pints, and it likewiſe
uſually vomits.

L E E Z, *near the Earl of Manchester's,
Essex.*

It is a chalybeate water, similar to
those of *Islington* and *Hampstead.*

L I N C O M B, *near Bath, in Somerset-
shire.*

This is a chalybeate and acidulous
water, containing natron, with a small
quantity of purging salt. It soon
loses its virtue if exposed to the air;
and in a few days also in bottles.

It resembles, in its virtues, the waters of *Thetford* and *Ilmington.* . .

L I S B E A K, *in the parish of Killasher,
in the county of Fermanagh, Ireland.*

Here are two strong sulphureous
waters, much of the same kind.

They yield upwards of thirty grains
of natron in the gallon, and it is more

free

free from heterogeneous mixtures than in moſt waters.

L I S-D O N E-V A R N A, *in the county of Clare, in Ireland.*

This is a ſtrong chalybeate water, and manifeſts itſelf as ſuch both to the taſte and ſmell. It is alſo impregnated with natron.

It keeps its virtue in well-corked bottles.

It uſually vomits and purges on firſt drinking, but afterwards operates by urine.

It ſeems to reſemble, in virtues, the *Thetford* and *Ilmington* waters.

L O A N S B U R Y, *in Lord Burlington's park, in Yorkſhire.*

This is a ſulphureous water, weakly impregnated with a purging ſalt.

It ſeems to be of the nature of the *Aſkeron* water; but is only uſed at pre-

ſent

fent for wafhing mangy dogs and fcabby horfes.

MACCROOMP, *in Ireland, about fix-teen miles from Cork.*

This is a chalybeate water, impregnated with natron, and refembles the *Thetford* and *Ilmington* waters.

MAHEREBERG, *fituated near Branden Bay, in the county of Kerry, Ireland.*

It is of the nature of the *Barrow-dale* water, but contains a much fmaller quantity of fea falt. The dofe for a purge is from a pint to a quart.

MALLOW, *in the county of Cork, Ireland.*

This is a warm water, perfectly limpid and pleafant-tafted, and keeps long in bottles well corked.

It

It is recommended in moſt caſes for which the *Briſtol* water has been uſed.

MALTON. *The Spring lies at the weſt end of the town of New Malton, in Yorkſhire.*

It is a ſtrong chalybeate, abounding with fixed air when freſh drawn; has a ſaltiſh taſte, and leaves a bitterneſs in the throat.

A gallon yields nearly two drams of vitriolated magneſia.

It operates by ſtool and urine. The doſe is from three pints to twice that quantity. If the ſtomach be foul, it is apt to vomit. In its virtues it re-ſembles the *Scarborough* water.

MALVERN, *in Glouceſterſhire.*

There are two noted ſprings at this place, one of them called the *Holy Well*, in the midway between Great and Little Malvern, the other is about

a quar-

a quarter of a mile from Great Malvern. · But the waters are not mate-
rially different.

They are light and pleafant chalybeates, and are remarkable for being
almoft entirely free from any earthy
matter; for three quarts of the Holy
Well water being evaporated, fcarce
the fourth part of a grain of fediment
was left behind.

They are recommended as excellent
in difeafes of the fkin; leprofies; fcorbutic complaints; the King's evil;
glandular obftructions; fcald heads;
old fores; cancers, &c. They are alfo
ferviceable in inflammations and other
difeafes of the eyes; in the gout and
ftone; in cachectic, bilious, and paralytic cafes; in old head-achs, and in
female obftructions.

The external ufe is by wafhing the
part under the fpout feveral times in a
day; afterwards covering the part with
cloths

cloths dipt in the water, which muſt be kept conſtantly moiſt. Thoſe who bathe, uſually go into the water with their linen on, and dreſs upon it wet, and it is never found to be attended with inconvenience.

The waters, when firſt drunk, are apt to occaſion, in ſome, a ſlight nauſea; others they purge briſkly for ſeveral days; but they operate by urine in all.

It is adviſeable to drink freely of the waters for ſome days before they are uſed externally.

MARKSHALL, *in Eſſex.*

This is a chalybeate water, reſembling thoſe of *Iſlington* and *Hampſtead.*

MATLOCK, *near Wirkſworth, in Derbyſhire.*

At this place (which is perfectly romantic) are ſeveral ſprings of warm water,

water, which appear to be of the nature of the *Briſtol* water, except that it is very ſlightly impregnated with iron.

Its heat is about 69°, and its virtues are ſimilar to thoſe of the *Briſtol* and *Buxton* waters.

The baths are recommended in rheumatic complaints, in cutaneous diſorders, and in other caſes where warm bathing is ſerviceable.

There are great numbers of petrifactions in the courſe of this water.

MAUDSLEY, *near Preſton, in Lancaſhire.*

The water is of a blueiſh colour, has a fœtid ſmell, and a brackiſh taſte.

It is a ſtrong ſulphureous water, and contains about two ounces of ſea ſalt in the gallon.

It is purgative, and has nearly the ſame virtues as the *Harrogate* water.

MECHAN,

MECHAN, *in the county of Ferma-*
nagh, Ireland.

In this place there are two fulphureous fprings, both of the fame nature.

They contain natron, and in their
virtues refemble the *Drumgoon* and
Swadlingbar waters.

MILLAR's SPA, *Stockport, in the*
county of Lancafter,

It is a chalybeate water of the nature of that of *Tunbridge*, but feems
to be ftronger of the iron.

MOFFAT, *in the county of Annandale,*
Scotland.

At this place there are two fprings
or wells.

They are both fulphureous, and
have a ftrong fmell and tafte; the upper one is the ftrongeft, and moft naufeous, and lefs drunk of than the other,
though as it bears heat better it is moft
ufed for bathing.

The

The Moffat water is alterative, and diuretic, but it fometimes purges.

From a gallon were obtained three grains of earth, and fifty grains of marine falt, though probably mixed with a fmall quantity of vitriolated magnefia.

It being fufpected by Dr. Plummer to contain copper, the Rev. Dr. Walker put a polifhed plate of iron into the well, and he found after fome time it had contracted a green ruft. This, in his opinion, confirms Dr. Plummer's conjecture.

MORETON, or MORETON-SEE, *fituated about two miles weft of Market-Drayton, in Shropfhire.*

It is efteemed as an excellent cooling and diuretic purge. It operates brifkly; is pungent to the tafte, and feems to be of the nature of *Holt* water.

The gallon contains 200 grains of vitri-

vitriolated magnefia, and 76 of calcareous earth.

Moss House, *near Maudfley, in Lancafhire.*

This is a brifk chalybeate water, and in its virtues refembles thofe of *Hampftead* and *Iflington.*

Mount D'or, *feven leagues from Clermont, in the Auvergne, France.*

The water is warm, and of the nature of the *Aix-la-Chapelle.*

Taken internally it acts as a diuretic, and it fometimes purges. Bathing in it fweats profufely, without weakening the patient.

It has been recommended in the rheumatifm, gout, palfy, and many other diforders.

Mount Pallas, *in the county of Cavan, Ireland.*

It is a chalybeate water, and feems to be of the nature of the *Athlone.*

NEVIL-

NEVIL-HOLT, *near Market-Harborough, in Leicestershire.*

This is an exceeding clear water as it falls from the spout, and is void of all smell. It has a brisk, austere, bitter, yet not disagreeable, taste, and abounds in fixed air.

Exposed to the air, it soon becomes turbid, and spoils. But in well-closed bottles it will keep good.

A gallon of the water contains two drams of vitriolated magnesia, two drams eighteen grains of muriated argillaceous earth, and eighteen grains of muriated magnesia.

Drunk to the quantity of several pints, it proves purgative, and operates without griping.

It also operates by urine and sweat.

It is a powerful antiseptic in putrid diseases.

When taken as an alterative, it must

be

be taken in fmall dofes, from a few fpoonfuls to a quarter or half a pint, feveral times in a day, according to its effect; and a little brandy, or the like, may be added if it fit cold on the ftomach.

It is efteemed an excellent remedy in old dyfenteries and diarrhœas, in internal hæmorrhages, in the fluor albus, and gleets, in the gravel, in rheumatifms, and for the worms; it is good in atrophies, in bloated conftitutions, and dropfical complaints, in fcorbutic diforders, in want of appetite, and in other cafes; in inflammatory complaints however, and where there is an acidity of the humours, it does mifchief.

Externally, it is a fpeedy cure for frefh wounds, for inflamed eyes, and hectic ulcers, &c. efpecially if taken inwardly at the fame time.

NEW

New Cartmal.

See *Rougham*.

Newnham Regis, *in Warwick-*
shire.

There are three wells at this place: they are all of them chalybeate, laxative, and diuretic; and feem to refemble the *Scarborough* water.

They have fomewhat of a fulphureous fmell.

Newton Dale, *in the North Riding of Yorkshire.*

This is a cold petrifying water.

It is faid to cure effectually loofeneffes, and bleedings of every kind; and that it reftores weakened joints, though beginning to be diftorted, by bathing in it.

New-

NEWTON STEWART, *near Caf-tlehill, in the county of Tyrone, Ireland.*

It is a chalybeate water, of the nature of that of *Tunbridge.*

NEZDENICE, *in Germany, about half a mile from Hunnobroda, in the diftrict of the caftle of Banow. The Spring is near this village.*

This is an acidulous water, impregnated with fixed air like thofe of *Seltzer* and *Pyrmont.*

It is in great repute among the people in the neighbourhood, for the cure of many diforders, particularly thofe in which the waters juft mentioned are ferviceable.

NOBBER, *in the county of Meath, Ireland.*

It is a vitriolic water, and refembles thofe of *Hartfell* and *Crofs-town.*

L   NOR-

NORMANBY, *four miles from Pickering, in Yorkshire.*

It is clear, beautiful, and fœtid, and when poured out fparkles like Champagne.

It is a fulphureous, and gently purgative water, and refembles the *Afkeron* water in virtues. A gallon yields fcarcely. twenty grains of vitriolated magnefia, and about half that quantity of fea falt.

Near it is a chalybeate water, called *Nether Normanby Spa*; a gallon of which afforded ten grains of fea falt.

NORTH-HALL.

See *Barnet.*

NOTTINGTON, *near Weymouth, in Dorfetfhire.*

This is a ftrong fulphureous water.

Its flavour refembles that of boiled eggs; and its colour, in a tin veffel,

is

is blue. A fhilling put into it at the fountain-head, becomes, in a few minutes, blue.

It contains 30 grains of natron, and feven of earth, in the gallon.

It is in repute for curing foulneffes of the fkin.

ORSTON, *in the county of Nottingham, near Thoroton.*

This water has a delicious, gentle, rough, fweetifh, chalybeate tafte, and a flight fulphureous fmell. It is replete with fixed air, fparkles and flies when poured out into a glafs, and makes the heads of thofe who drink it giddy.

It foon fpoils by expofure to air.

It is purgative, and feems to be poffeffed of the fame virtues as the *Pyrmont* water, for which it may be ufed as a fubftitute.

O U L T O N, *in Norfolk.*

It is a flight chalybeate water, fimilar to that of *Iflington.*

O W E N B R E U N, *in the county of*
Cavan, Ireland.

'This is a fulphureous water, impregnated with a purging falt, and a
little natron.

Its virtues refemble thofe of the
Afkeron water.

P A N C R A S, *in Middlefex,* *near*
London.

The water is almoft infipid to the
tafte.

It is impregnated with a purging
falt, together with a fmall portion of
fea falt.

It is therefore a purgative water,
and is alfo diuretic.

Its

Its virtues are allied to thofe of the *Cheltenham* water, and it is alfo of fervice in the ftone, gravel, and fimilar diforders.

Passy, *near Paris, in France.*

It is a clear, colourlefs, chalybeate water, with a fubacid tafte, and ferruginous fmell, and emits plenty of air-bubbles.

It is a ftrong chalybeate water, but does not prove purgative, unlefs drunk in large quantity. It is of the nature of *Pyrmont* water.

Peterhead, *in the county of Aberdeen, Scotland.*

This is one of the ftrongeft, and moft famous chalybeate waters in Scotland. It is of the nature of our *Iflington* water, but more powerful.

Pet-

PETTIGOE, *in the county of Don-negal, Ireland.*

It is one of the ſtrongeſt ſulphureous waters in Ireland; and is impreg-nated with vitriolated magneſia, of which it contains near 50 grains in the gallon.

In its virtues it reſembles the *Aſke-ron* water.

PISA, *in Italy.*

About 16 miles from Piſa is a warm bath called *Bagno a Acqua,* and at the bottom of Mount Piſa, now called St. Julian, 12 miles from the town, are a number of ſprings of warm water, uſed both for drinking and bathing.

The hotteſt raiſes Fahrenheit's ther-mometer to 104°; the cooleſt to 92°.

In ſmell and taſte they differ not
from

from common water. They contain natron, fea falt, and felenite.

Thefe waters are diaphoretic and diuretic, and, if drunk in large quantity, often operate by ftool.

PLOMBIERES, *in Lorraine, France.*

The water is tepid and faponaceous, with a faltifh tafte.

It is recommended for complaints of the ftomach proceeding from acidity; in fpitting of blood; in hæmorrhages; phthifical and afthmatic complaints; in dropfy of the belly; the diabetes; fluor albus; dyfentery; and in all cutaneous diforders.

It is drunk from a pint to three quarts, on an empty ftomach, in the morning; it is diuretic and laxative.

It is alfo ufed outwardly as a bath; and is reckoned excellent for wafhing ulcers.

PONTGIBAUT, *in Auvergne, France.*

The water is limpid, fubacid, and contains about 55 grains of natron, and 50 of calcareous earth, in the gallon.

It is diuretic and gently opening; and its virtues are allied to thofe of the *Tilbury* and *Seltzer* waters.

PYRMONT, *in Weftphalia, Germany.*

This is a very brifk chalybeate, abounding in fixed air; and when taken up from the fountain, fparkles like the brifkeft Champaign wine. It has a fine, pleafant, vinous tafte, and a fomewhat fulphureous fmell. It is perfectly clear, and bears carriage better than the *Spa* water.

A gal-

A gallon of it contains 46 grains of chalk, 15.6 of magnefia, 30 of vitriolated magnefia, 10 of fea falt, and 2.6 of aerated iron *.

Perfons who drink it at the well are affected with a kind of giddinefs or intoxication; owing, it may be fuppofed, to the great quantity of fixed air with which the water abounds.

The common operation of this water is by urine; but it is alfo gently fudorific; and if taken in large quantity proves laxative. When, however, it is required to have this latter effect, it is ufual to mix fome falts with the firft glaffes.

It is drunk by glafsfuls in the morning, to the quantity of from one to five or fix pints, according to cir-

* Dr. Marcard, in his *Defcription of Pyrmont*, on the authority of M. Weftrumb of Hammeln, eftimates the iron at fomewhat more than eight grains to the gallon.

cumftances,

cumſtances, walking about between each glaſs.

It is recommended in caſes where the conſtitution is relaxed; in want of appetite and digeſtion; in weakneſs of the ſtomach, and in heartburn; in the green ſickneſs; in female obſtructions, and in barrenneſs; in the ſcurvy, and cutaneous diſeaſes; in the gout, eſpecially when mixed with milk; in cholics; in bloody fluxes; in diſeaſes of the breaſt and lungs, in which caſes it is beſt taken lukewarm; in nervous, hyſteric, and hypochondriacal diſorders; in apoplexies and pàlſies; in the gravel, and urinary obſtructions; in foulneſs of the blood; and in obſtructions of the finer veſſels. It amends the lax texture of the blood; exhilarates the ſpirits without inflaming, as vinous liquors are apt to do; and is among the beſt reſtoratives in decayed and broken conſtitutions.

This

This water poſſeſſes the general virtues of the *Spa* water; and at the fountain it is even more ſpirity, as well as a ſtronger chalybeate. The reader therefore is referred to what is ſaid of the *Spa* water, for a further account of its virtues.

To thoſe to whom œconomy is an object it may be of importance to know, that the expence of living at Pyrmont is not above half what would be incurred at Spa.

QUEEN CAMEL, *near Wincaunton, in Somerſetſhire.*

The water has a fœtid, ſulphureous ſmell, like the waſhings of a foul gun. It tinges ſilver of a yellow or black colour, and blackens the ſtones on which it runs. It is alſo ſaid to be colder than common water.

It contains natron, together with

ſea

fea falt, a chalky earth, and a bituminous or fulphureous matter.

It has been ufed with fuccefs both inwardly and outwardly in cutaneous diforders, the fcurvy, and the King's evil; and for thefe purpofes a place is contrived for bathing.

RICHMOND, *in the county of Surry.*

This is a purging water, of the nature of thofe of *Acton* and *Pancras.*

RIPPON, *in Yorkfhire.*

Near this place a fpring of a pretty ftrong fulphureous water rifes from a limeftone hill. A gallon yielded, on evaporation, 66 grains, of which nearly half was earth, the remainder fea falt.

ROAD, *in Wiltfhire.*

This is a chalybeate water with a fulphureous fmell, and is impregnated with natron.

It

It is recommended internally and externally in fcorbutic and fcrophulous cafes, and in cutaneous difeafes, &c.

On firft taking this water it acts as a gentle purge.

It does not bear carriage.

ROUGHAM, *in Lancafhire.*

It is of the nature of *Barrowdale* water, but much weaker.

The gallon contains five drams of fea falt, and one dram of vitriolated magnefia.

The dofe for a purge is from three to eight quarts. In that quantity it operates powerfully by ftool, and alfo by urine.

SAINT AMAND, *a town in French Flanders.*

There are two fountains here, one called *Bouillon* or *Bouillant,* the other

the

the fountain of *Arras*, or *L'Eveque
d'Arras*, the latter of which is the
ftrongeft.

They fomewhat refemble thofe of
Aix-la-Chapelle in appearance, but are
inferior in heat, raifing Fahrenheit's
thermometer to 75° only, when in the
open air it ftood at 50°.

They principally deferve notice on
account of the *boue* or mud baths.
The method of ufing them is to bury
the affected limb, or part of the body,
even up to the armpits, for fome hours,
as the cafe may require : the patient is
then carried to a hot bath and cleanf-
ed from the black mud which adheres
to the fkin.   The *boue* is of fo firm
a confiftence that a part muft firft be
digged out. A thermometer immerfed a
foot deep in it was raifed to near 60°,
when in the open air it was at 47°. It
contains lime, magnefia and iron, all
aerared,

aerated, befides felenite, argillaceous and filiceous earth. The refiduum of the water exhibits the fame ingredients.

They both contain a peculiar air, in fmell very much refembling hepatic, which Dr. Afh attributes to a bituminous fubftance.

SAINT BARTHOLOMEW's WELL, *Ireland. It is about two miles fouth-weft from Cork.*

The water is foft, and mixes fmoothly with foap.

By keeping it putrifies, and then tinges filver, and throws up a ftinking fcum which has fomewhat of an irony tafte. Galls then give it a purple tinge, which they do not to the frefh water.

The gallon affords 24 grains of refiduum, which is chiefly natron.

Its

Its virtues are fimilar to thofe of the *Tilbury* water.

SAINT ERASMUS'S WELL, *fituated on Lord Chetwynd's grounds in Staffordfhire.*

The water is of the nature of *Barrowdale*, but much weaker, the gallon yielding only four drams 32 grains of fea falt.

It is of the colour of fack, but without much tafte or fmell.

If drunk to the quantity of feveral quarts, it operates powerfully by ftool.

SALES SPA.

See *Lancafter*.

SCARBOROUGH, *in Yorkfhire*.

The waters of this place are chalybeate and purging; and they are more frequented and ufed than any other water of this clafs in England.

There

There are two wells; the one more purgative, the other a ftronger chalybeate. Hence the latter (which is nearcft the town) has been called the *chalybeate* fpring, the other the *purging*; though they are both impregnated with the fame principles, but in different proportions. The *purging* is the moft famed, and is that which is ufually called the *Scarborough* water. This contains 52 grains of calcareous earth, two of ochre, and 266 of vitriolated magnefia, in the gallon: the chalybeate, 70 grains of calcareous earth, 139 of vitriolated magnefia, and 11 of fea falt.

When thefe waters are poured out of one glafs into another, they throw up a number of air-bubbles; and if fhaken for a while in a clofe ftopt phial, and the phial be fuddenly opened before the commotion ceafes, they dif-

plode

plode an elaftic vapour with an audible noife, which fhows that they abound in fixed air.

At the fountain they both have a brifk, pungent, chalybeate tafte; but the *purging* water taftes bitterifh, which is not ufually the cafe with the *chalybeate* one.

They lofe their chalybeate virtues by expofure, and alfo by keeping; but the purging water fooneft.

They both putrify by keeping; but in time recover their fweetnefs.

Four or five half pints of the *purging* water drunk within an hour, give two or three eafy motions, and raife the fpirits. The like quantity of the *chalybeate* purges lefs, but exhilarates more, and paffes off chiefly by urine.

Thefe waters have been found of fervice in hectic fevers, in weakneffes of the ftomach, and indigeftion; in

relaxations

relaxations of the fyftem; in nervous, hyfteric, and hypochondriacal difor-ders; in the green ficknefs, in the fcurvy, rheumatifm, and afthmatic complaints; in gleets, the fluor albus, and other preternatural evacuations, and in habitual coftivenefs. The wa-ters are to be varied according to the intention to be anfwered.

SCOLLIENSES, *in Upper Rhoetia, Switzerland.*

It is a chalybeate water, impreg-nated with natron; and fo full of fixed air, that it often burfts the bottles in which it is kept.

It makes the drinkers giddy, and operates mildly, though largely, by ftool, and by fpitting.

It is efteemed excellent for cholicy pains, both as a cure and preven-tative.

In

In its general virtues it refembles the *Spa* water.

Sea Water.

Sea water has a falt, bitterifh tafte, appears of a greenifh colour, and becomes fœtid by keeping.

As an immenfe number of fprings, rivers, &c. are continually emptying themfelves into the fea, as it contains en almoft infinity of animals and vegetables, and covers and wafhes fuch a variety of lands and fhores, it would feem to be impregnated with very heterogeneous matters. Neverthelefs, the water, in different parts of the ocean, appears to be nearly alike, and the difference in its contents to be much lefs than might at firft be imagined.

A gallon, taken up off Brighthelm-ftone, 400 yards from low-water mark,
yielded

yielded three ounces 323$\frac{1}{4}$ grains of
fea falt, one ounce 283$\frac{3}{4}$ grains of muriated magnefia, and 93$\frac{1}{4}$ grains of
gypfum. It alfo afforded one ounce
meafure of fixed air, four of atmofpheric, and one of phlogifticated.

Sea water, in hotter climates, generally contains a greater proportion of
thefe matters than that in colder ones,
and therefore is ftronger. The difference, in fome places, is above two
to one.

Sea water taken internally, in a fmall
quantity, proves a ftimulating and
heating remedy, diffipating the finer
fluids, and occafioning thirft.

In a larger quantity it proves purgative. But differs from other purges
in that patients who drink it daily for
a confiderable time, inftead of lofing,
often gain ftrength by it.

It therefore acts not merely as a
purgative, but gives alfo a brifk ftimulus

mulus to the ftomach and inteftines, thereby increaſing the appetite, and promoting digeftion.

By means of this excellent property of fea water (viz. our being able to keep up a purging for a conſiderable time, without hurting the conftitution) we are enabled frequently to remove diforders which have refifted the force of other remedies.

It is of excellent uſe in fcrophulous complaints; and glandular fwellings are generally removed by it. If joined with the bark, it has fometimes a better effect in thofe cafes.

It is alfo ferviceable in purging off grofs humours, which have been the confcquence of indulging the appetite too freely, and leading an inactive life: alfo in cleanfing the inteftines of viſcid mucus, and worms.

In cafes where there is fever, heat, or inflammation, fea water is found to be be hurtful. Previous to its ufe, therefore, thefe fymptoms fhould be removed by bleeding, purging, and a proper cooling treatment.

As fea water is fpecifically heavier than common water, and (by reafon of the faline matters contained in it) is alfo more ftimulating, it is more efficacious when ufed externally as a bath.

It has alfo particular virtues when externally ufed. On account of its ftimulating and difcutient property, it is excellent in the fcrophula or king's evil, in hard fwellings, in the bite of a mad dog, in the rickets, in the dry leprofy and itch,- in paralytic and fcorbutic complaints, and in many other cafes. But in moft of thefe, it is proper to ufe it both internally and externally.

SED-

S E D L I T Z, *Germany, a village in Bohemia.*

This purging water is of the fame nature as our Epfom, but much ftronger, a gallon yielding about two ounces of the purging falt.

Two or three tea-cups full are generally fufficient for a dofe; and the ftrongeft conftitution fcarce requires more than a pint.

S E L T Z E R, *in Germany. This Spring is near to the town of Neider, or Lower Seltzer, about three leagues from Franckfort on the Maine, in the Lower Archbifhoprick of Treves.*

It rifes near a fmall trout ftream. The country and avenues around are delightful, and afford a very pleafing profpect.

The water iffues forth with great
rapidity;

rapidity ; is remarkably clear and light, and on pouring it from one veffel to another, plenty of air-bubbles arife.

It has, at firft, fomewhat of a brifk fubacid pungent tafte, but leaves behind a lixivial one.

If expofed twenty-four hours to the air, it lofes entirely its original tafte, and acquires that of a flat alkaline ley. But no fediment is depofited.

It putrifies fooner than any other medicinal water.

When frefh, it makes an immediate effervefcence with acids, but efpecially with Rhenifh wines, and a little powdered fugar.

It alfo curdles with a folution of foap.

It does not change purple with galls ; and therefore contains no chalybeate.

Oil of tartar dropt into it makes it milky, but does not occafion a precipitate.

M   It

It contains 14 grains of chalk, 20½ of magnefia, 141.6 of natron, and 92 of fea falt, in the gallon. From this quantity of the water 128 ounce meafures of fixed air were obtained.

Its operation is chiefly by urine, for it has no purgative virtues. It corrects acidities, renders the blood and juices more fluid, and promotes a brifk and free circulation. Hence it is good in obftructions of the glands, and againft grofs and vifcid humours.

It is of great ufe in the gravel and ftone, and in other diforders of the kidnies and bladder.

It is alfo excellent in gouty and rheumatic complaints*, efpecially when mixed with milk †.

* In thefe diforders its virtue is faid to be much improved by the addition of Rhenifh wine, and a little fugar.

† Affes, or goats milk, is ufually preferred.

It

It is drunk with great fuccefs in fcorbutic, cutaneous, and putrid difforders.

It is good againft the heart-burn ; and it is alfo an excellent ftomachic. Several pints may be drunk in the courfe of a day.

On account of its diuretic quality, it is of fervice in dropfical complaints.

Mixed with affes milk, it is of great ufe in confumptive cafes, and in diforders of the lungs.

It is in great efteem in nervous difforders, either with, or without milk, as is found to be moft fuitable to the conftitution.

It is alfo of fervice in hypochondriacal and hyfteric complaints, and in obftructions of the menfes, efpecially if exercife be ufed.

It is given in purgings and fluxes arifing from acidity in the bowels, with good fuccefs.

Drunk

Drunk by nurfes, it is faid to render their milk more wholefome and
nourifhing to children, and to prevent it from turning four on their
ftomachs.

As the fixed air of this water fo
foon flies off, it ought either to be
drunk on the fpot, or at leaft fhould
be impregnated with a frefh quantity
previous to its being taken, according
to the directions given in the beginning of this treatife.

Thofe perfons, with whofe ftomachs water alone does not fo well
agree, are advifed to mix with it fome
generous and agreeable wine, in cafes
where wine will not be hurtful. (See
alfo *Spa* and *Pyrmont* waters).

SENE, or SEND, *near the Devizes,*
Wiltfhire.

At this place are two chalybeate
fprings, one of them ftronger than
the

the other, but both refembling in virtues the *Hampftead* and *Iflington* waters.

They are diuretic, but not purgative.

At a village called *Paulfholt*, near
this place, is another chalybeate
fpring.

SEYDSCHUTZ, *in Germany.*

It is fituated near to that of *Sedlitz,*
and is of the fame purgative nature,
but fomewhat ftronger.

SHADWELL, *near London, fituated in Sun Tavern Fields.*

This is a vitriolic chalybeate water, and is one of the ftrongeft waters
of the kind in England; it alfo contains iron held in folution by aerial
acid. The gallon yielded 1132 grains

of

of martial vitriol, and 188 of an ochry-coloured earth.

It has an acid, auftere, vitriolic tafte, and with galls gives a blueifh black colour like ink.

It has been taken inwardly to the quantity of a pint, divided into two or three dofes in the courfe of an hour in the morning. It vomits, and gently purges; it turns the ftools black.

It has been found of 'fervice in the fluor albus, in obftinate gleets, and dyfenteries; in inward bleedings; in the jaundice; and in fcorbutic and leprous cafes. But it has chiefly been ufed externally for fore eyes, the itch, fcabs, tetters, fcald-head, ulcers, fiftulas, and the like, by wafhing, or elfe applying linen rags dipped in it to the parts.

In fcorbutic and leprous cafes, the internal ufe is firft advifed till the
eruptions

eruptions are thrown out, which are then to be removed by the outward application of the water.

SHAPMOOR. *The Spring is situated in a marshy heath, between Shap and Orton, in Westmoreland.*

This is a sulphureous water, impregnated with a purging salt, composed of vitriolated magnesia, sea salt, and natron, about 370 grains in the gallon.

Three pints of it prove purgative.

In its virtues it seems to resemble the *Askeron* water.

SHUTTLEWOOD, *situated between Lyver and Romeley, in Derbyshire.*

It is a sulphureous water, containing near two drams of sea salt in the gallon.

M 4 Its

Its virtues refemble thofe of the *Harrogate* water.

Shipton, *in Yorkſhire.*

It is a fulphureous water, impreg-nated with fea falt, together with a purging falt.

In its virtues it refembles the *Har-rogate* water.

Somersham, *in Huntingdonſhire.*

This is a chalybeate water, impreg-nated with green vitriol and alum, and contains alfo fixed air.

The feafon for drinking it is from May to October.

It is drunk in the morning to the quantity of feveral glaſſes. It is recom-mended in debilities of the ſtomach and bowels, in dyfenteries, hæmor-rhoids, and worms, in nidorous crudities, in obſtructions of the liver and fpleen,

in

in uterine complaints, in the ſtone
and gravel, in the ſcurvy, in hyſteric
and hypochondriacal diſorders, and
many others.

It is proper to purge before and
after the courſe, and ſalts may alſo be
occaſionally added to it.

Externally it is applied to foul ulcers and cancers.

S P A, *in the biſhoprick of Liege, Germany, twenty-one miles ſouth-eaſt
from the town of Liege.*

In and about this town there are
ſeveral ſprings, which afford excellent chalybeate waters: and in Great
Britain they are the moſt drunk of
any foreign mineral waters.

The principal ſprings are,

1. The POHOUN, or POUHON, ſituated
    in the middle of the village.

2. SAUVINIERE, about a mile and a
    half eaſt from it.

3. GROIS-

3. GROISBEECK, near to the Sauvi-
    niere.
4. TONNELET, a little to the left of
    the road to the Sauviniere.
5. WATROZ, near to the Tonnelet.
6. GERONSTERE, two miles fouth
    of the Spa.
7. SARTS, or NIVERSET, in the dif-
    trict of Sarts.
8. CHEVRON, or BRU, in the princi-
    pality of Stavelot.

  9. COUVE,
10. BEVERSEE, } All near Malmdy.
11. SIGE,
12. GEROMONT.

The POUHON is a flow deep fpring,
and is more or lefs ftrong or gafeous
according to the ftate of the atmofphere.

The gallon contains 10 grains of
chalk, 30 of magnefia, 10 of natron,
and five of aerated iron. It yields of
fixed air 132 ounce meafures.

                              It

It contains more iron than either of the other fprings, and does not fo foon lofe its gas.

It is in its moft perfect and natural ftate in cold, dry weather. It then appears colourlefs, tranfparent, and without fmell, and has a fubacid chalybeate tafte, with an agreeable fmartnefs: at fuch times, if it be taken out of the well in a glafs, it does not fparkle; but after ftanding awhile, covers the glafs on the infide with fmall air-bubbles; but if it be fhaken, or poured out of one glafs into another, it then fparkles, and difcharges a great number of air-bubbles at the furface.

In warm, moift weather, it lofes its tranfparency, appears turbid or wheyifh, contains lefs fixed air, and is partly decompofed. A murmuring

M 6 noife

noife alfo is fometimes heard in the well.

It is colder than the heat of the at-mofphere by many degrees.

It is fuppofed to contain the great-eft quantity of fixed air of almoft any acidulous water; and in confequence thereof has a remarkable fprightli-nefs and vinofity, and boils by mere warmth. This, however, foon flies off, if the water be left expofed; though in well corked bottles it is in a great meafure preferved.

It is capable of diffolving more iron than it naturally contains, and there-by becoming a ftronger chalybeate. This is owing to the great quantity of fixed air which it contains.

For the fame reafon an ebullition is raifed in this water on the addition of acids, as they difengage its fixed air.

It

It mixes fmoothly with milk, whether it be cold or of a boiling heat.

Of the SAUVINIERE water, a gallon yields 6.5 grains of chalk, 4.5 of magnefia, two of natron, 3.5 of kali, 2.2 of aerated iron, and 108 ounce meafures of fixed air.

At the well it has fomewhat a fmell of fulphur.

GROISBEECK. The water is of the fame nature as the Sauviniere, but contains a fomewhat larger proportion of the feveral ingredients. It has a vitriolic tafte, and fomewhat of a fulphureous fmell.

TONNELET. This is one of the moft fprightly waters in the world. It is much colder than either of the other Spa waters; has no fmell; is bright, tranfparent, and colourlefs; and from the rapidity of its motion does not foul its bafon. It has a fmart, fubacid,

fprightly

fprightly tafte, not unlike the brifkeft Champaign wine.

From a variety of experiments it appears, that this water is more ftrong-ly charged than any of the others with fixed air, on which the *energy* of all waters of this kind depends, but it parts with it more readily.

It contains more iron than either of the fprings, except the Pouhon.

WATROZ. Its fituation is loweft of any of the fprings about Spa, and it is more apt to be foul: but when the well is cleaned out, and the water pure, it is found to be of the fame nature as that of POUHON. It is not purgative, as fome have afferted.

GERONSTERE. This water has much lefs fixed air than the POUHON. It has a fulphureous fmell at the foun-tain, which it lofes by being carried to a diftance. This fmell is ftrongeft in warm moift weather.

The

The air, or vapour, of this water affects the heads of some who drink it, occasioning a giddiness, or kind of drunkenness, which goes off in a quarter or half an hour. The Pyrmont, and several other brisk chalybeate waters, are found to have the same effect.

It is colder than any of the springs, the *Tonnelet* excepted.

SARTS, or NIVERSET. It resembles the *Tonnelet* water, but is rather less brisk and gaseous. It is however more acid and styptic.

BRU, or CHEVRON. The physicians at Liege have artfully decried this water, because it is not in the principality of Liege. But by every trial it appears not much inferior to any of the *Spa* waters. In the quantity of fixed air and of iron it contains, it approaches the Pouhon.

COUVE and BEVERSEE. The *Couve* nearly resembles the *Tonnelet* water;

4   or

or rather, may be placed in a medium between that and the *Watroz*. It hardly equals the tranfparency, fmart-nefs, and generous vinous tafte of the firft, but it greatly furpaffes the latter. The *Beverfce* agrees with this, only that it does not retain its fmartnefs fo well by keeping.

La Sige. It has fome of the general properties of the *Spa* waters, but in other refpects it is different.

It is moderately fubacid, fmart, and grateful, but has no fenfible chalyb-eate tafte.

It fparkles like Champaign wine when poured from one glafs to an-other. Upon ftanding it lofes its fixed air, and throws up a thick mo-ther-of-pearl coloured pellicle.

It is much more loaded with earthy matters, and lefs impregnated with iron and fixed air, than the other Spa waters.

Geromont. As a chalybeate and
acidu-

acidulous water it feems to be nearly
of the fame ftrength with *La Sige*;
but it contains a greater quantity of
natron, together with a mixture of
fea falt. The earthy matters, however,
are lefs.

Their Virtues, &c.—It appears,
that thefe waters are compounded of
nearly the fame principles, though in
different proportions. All of them
abound with *fixed air*. They contain
more or lefs iron, natron, and calcareous and felenitical earths; together
with a fmall portion of fea falt, and
an oily matter common to all waters.
Thefe are all kept fufpended, and in
a neutral ftate, by means of the aerial
acid, or fixed air.

From a review of the contents of
thefe waters, it cannot be imagined
that their virtues principally depend
on the fmall quantity of *folid* matters which they contain. They muft
there-

therefore depend moftly on their fubtle mineral fpirit, or *fixed air*. And they are probably rendered more active and penetrating both in the firft paffages, and alfo when they enter the circulation, by means of that fmall portion of iron, earth, falt, &c. with which they are impregnated.

Thefe waters are diuretic, and fometimes purgative; like other chalybeate waters they tinge the ftools black.

They exhilarate and affect the fpirits with a much more kind and benign influence than wine or fpirituous liquors; and their general operation is by ftrengthening the fibres. They cool and quench thirft much better than common water.

They are therefore found excellent in cafes of univerfal languor or weaknefs, proceeding from a relaxation of the ftomach, and of the fibres in general, and where the conftitution has
been

been weakened by difeafe, or by too
fedentary a life. In weak, relaxed,
grofs habits; in nervous diforders; in
the end of the gout and rheumatifm,
where the conftitution needs to be re-
paired; in fuch afthmatic diforders and
chronic coughs as proceed from too.
great a relaxation of the pulmonary
veffels; in obftructions of the liver
and fpleen; in cafes where the blood
is too thin and putrefcent, occafioned
by irregularities, or by fcorbutic or
other putrid diforders; in hyfterical
and hypochondriacal complaints, where
the fibres are too irritable and relaxed, and where the habit in general
needs to be reftored; in paralytic diforders; in gleets; in the fluor albus;
in fluxes of the belly; and in other
inordinate difcharges proceednig from
too great weaknefs or relaxation of
any particular part; in the gravel and
ftone; in female obftructions; in barrennefs;

rennefs; and in moft other cafes where
a ftrengthening and brifk ftimulating
refolving chalybeate remedy is wanted; and where there are no confirmed
obftructions, or fo much heat and fever as to forbid their ufe.

They are, however, generally hurtful in hot, bilious, and plethoric conftitutions, when ufed before the body
is cooled by proper evacuations. They
are alfo hurtful in cafes of fever and
heat; in hectic fevers and ulcerations
of the lungs, and of other internal
parts, particularly where there is no
free outlet to the matter; and alfo in
moft confirmed obftructions attended
with fever.

The ufual feafon for drinking them
is in July and Auguft, or during the
fummer months from May to September. The water, however, is beft
in the winter, and in dry, frofty weather;

ther ; and probably might then be drunk to greateſt advantage.

If they lie cold on the ſtomach, a few carraway ſeeds, cardamoms, or other aromatic, may be taken with them. The veſſel out of which it is drunk may alſo be warmed with hot water, or a little warm water may be added immediately before drinking. It muſt always be drunk before noon.

The quantity to be drunk ſhould be different according to the age, conſtitution, and other circumſtances of the patient. The only certain rule is, that quantity which the ſtomach can bear without heavineſs or uneaſineſs. The greater the quantity any one drinks, the better, provided it agrees, and paſſes well off. It is adviſeable to begin with drinking a glaſs or two ſeveral times in a day, increaſing the quantity daily, as far as the ſtomach will bear. To continue that doſe during

ring the course, and to finish by lef-
fening it by the fame degrees by which
it was augmented. Moderate exercife is proper after drinking. It is to
be continued for feveral weeks or
months, according to the circumftances.

Previous to the ufe of the water, it
is proper to cleanfe the firft paffages
by gentle purges, and if judged neceffary, an emetic alfo fhould be given. During the courfe, likewife,
coftivenefs fhould be prevented, by occafionally adding Rochelle falts, or
rhubarb, to the firft glaffes of water
in the morning.

When there is too much heat, the
faline draughts, nitre, vegetable acids,
or the like, fhould be given; and the
elixir of vitriol has been added to the
water, in intermittent feverifh complaints, with good effect.

A cooling regimen fhould be obferved

ed while drinking thefe waters, as alfo regular hours, and quietnefs, or chear-fulnefs of mind.

In cafes of rigidity of the fibres, the warm bath is recommended, among the beft preparatives to a courfe of thefe waters; and, hence bathing at *Aix-la-Chapelle*, or at *Chaude Fontaine*, is often premifed. The warm bathing may occafionally be repeated during the courfe. In oppofite cafes, the cold bath is recommended.

The Spa water is ufed alfo externally, in a variety of cafes, with good fuccefs. It is ufed as an injection in the fluor albus, and in ulcers and cancers of the womb, and alfo in the gonorrhœa; it is ferviceable in venereal aphthæ, and ulcers in the mouth; it is ufed to wafh phagedenic ulcers; it is recommended by way of gargle for relaxed tonfils, and for faftening loofe teeth; it is alfo good in other relaxations;

tions; and it is faid to cure the itch, and fimilar complaints, by wafhing and bathing, an internal courfe being alfo obferved at the time.

As the Spa waters are impregnated with different proportions of the fame ingredients, they may be chofen differently, according to the intentions we have in view. The *Pohoun* is the ftrongeft chalybeate. The *Tonnelet* is a weaker chalybeate, but brifker, and rather more gafeous. The *Groifbeeck* and *Sauviniere* are ftill weaker chalybeates, but contain a portion of kali, which the others do not. The *Geromont* is likewife a weak chalybeate, but contains a great deal of calcareous and felenitical earth, and about three times as much alkaline falt as any of the others. The four laft waters, therefore, will be better in diforders arifing from an acid caufe, and as diuretics, particularly the *Geromont*.

S T A N-

STANGER, *near Cockermouth, in Cumberland.*

This is a falt chalybeate, or vitriolic water; and, when drunk to four or five pints, operates with violence both upwards and downwards.

STENFIELD, *in Lincolnfhire.*

It is a chalybeate laxative water, and refembles that of *Orfton.* It is light, clear, pleafant tafted, and full of gas at firft, but on long ftanding in its large refervoir fpoils.

STREATHAM, *in Surry, near London.*

The water has a yellowifh tinge, and throws up a fcum variegated with blue, green, and yellow. Its tafte is fomewhat faline and difagreeable.

N   The

The gallon contains 160 grains of falt compofed of fea falt and vitri-olated magnefia, and 40 of calcareous earth.

It is a mild purging water, and may be drunk to the quantity of three or four pints.

It is alfo diuretic, and is faid to be found ufeful in diforders of the eyes.

S T O K E.

See *Jeffop's Well*.

S U C H A L O Z A, *about a mile from Hungarian Broda, in Germany.*

It is an acidulous water, refembling that of *Nezdenice* in virtues.

It is greatly efteemed in the neigh-bourhood for the cure of fcrophulous and other diforders, in which waters of this kind are ferviceable; and is

drunk

drunk with victuals inftead of fmall beer and wine.

S u t t o n B o g, *in the county of Ox-ford, near to Northamptonfhire.*

This is one of the waters termed *fulphureous.*

It has an intolerable fœtid fmell, like rotten eggs. Its tafte is faltifh and pungent, like foap lees.

It throws up a blue fcum, and the mud at the bottom is jet black. In half an hour it turns filver of a copper colour.

It contains 131 grains of natron mixed with a little fea falt, and nine grains of argillaceous earth, in the gallon.

It is a mild laxative, or purging water.

It is ufed both for drinking and bathing; and ulcers, tumours, fcro-

N 2   phulous,

phulous, and other difeafes of the ſkin, are fuccefsfully waſhed with it. The mud is alſo made uſe of.

SWADLINGBAR, *in the county of Cavan, Ireland.*

The water is ſometimes tranſparent and colourleſs; at other times ſomewhat whitiſh.

It has a ſtrong ſulphureous ſmell, which it retains long in bottles well corked. It tinges ſilver of a blackiſh or copperiſh colour.

The well is commonly covered with a whitiſh or blueiſh ſcum; and depoſits a mud which burns, on a red hot iron, with a blue flame.

It contains natron, together with a little vitriolated magneſia and earth.

SWANSEY, *in Glamorganſhire, North Wales.*

It is impregnated with vitriolated iron,

iron, of which a gallon yields thirty-
two grains.

Dr. Rutty fufpects it to contain copper.

Taken inwardly it is alfo faid to ftop purgings; applied outwardly it ftops bleeding.

S Y D E N H A M, *in Kent, near London.*

The water is fomewhat bitterifh to the tafte.

It is purgative, and of the nature of *Epfom* water, but only about half the ftrength of it.

T A R L E T O N, *eight miles from Pref-ton, in Lancafhire.*

This is a chalybeate water, and drunk to the quantity of three or four pints proves purgative. In its virtues it feems to refemble the *Scarborough* water.

It

It has a fomewhat fulphureous fmell when firft drawn.

TEWKESBURY, *in Glouceſterſhire.*

It is a purging water, of the nature of thofe of *Acton*, *Pancras*, and *Ep-ſom*.

There are two other fprings of the fame kind in the neighbourhood; one of them is in Walton grounds*, the other in Teddington grounds.

THETFORD, *in the county of Norfolk.*

This is a chalybeate and acidulous water, and contains alfo natron.

It operates by urine, and alfo gently by ftool.

It is recommended in pains of the ftomach and bowels; in lofs of appetite; in relaxed ftate of the fibres; in

* See *Walton.*

hyfteric

hyfteric diforders; and in beginning confumptions.

THOROTON, *near Newark upon Trent, Nottinghamfhire.*

It is a chalybeate laxative water, refembling that of *Orfton.*

THURSK, *in the North Riding of Yorkfhire.*

It is a brifk, fparkling, chalybeate water, and is alfo purgative and diuretic. It refembles the *Scarborough* and *Cheltenham* waters.

TIBSHELF, *in Derbyfhire.*

This is a fine clear chalybeate; and when poured from one 'glafs to another, fparkles like the *Spa* water, which it refembles in virtues.

N 4 TIL-

TILBURY. *The Spring which affords this water is situated near a farm-house at West-Tilbury, near Tilbury-Fort, in Essex.*

This water is not quite limpid at the well, but is somewhat straw-coloured.

It is soft and smooth to the taste; though after being agitated in the mouth, it leaves a small degree of roughness on the tongue.

It throws up a scum variegated with several colours, which feels greasy, and effervesces with vitriolic acid.

It mixes smooth with milk, but curdles with soap. When boiled it turns milky; a fourth part of mountain wine fines it immediately, and all acids do the same.

A gallon of the water contains 37
grains

grains of chalk, 49 of true nitre, 82 of fea falt, and $1\frac{1}{2}$ of natron.

It operates chiefly by urine, though it is alfo fomewhat purgative; and increafes perfpiration.

It is in efteem for removing glandular obftructions; it is good in bloody fluxes, purgings, and the like; in diforders of the ftomach arifing from acidity; in the gravel; fluor albus; and immoderate flux of the menfes.

As a diuretic it is good in dropfical complaints.

It gently warms the ftomach, ftrengthens the appetite, and promotes digeftion; it is alfo of fervice in lownefs of fpirits. From its efficacy in removing obftructions of the glands, it is recommended in fcurvies and cutaneous difeafes; and its virtues in thefe complaints feem to be confirmed by the tingling which it occafions in the fkin.

The dofe is ufually a quart in a day.

T O B E R.

Tober Bony, *in Ireland.*

This fpring is fituated about four miles north of Dublin.

The water is fweet, and foon lathers with foap.

Before rain and wind it yields a fœtid fmell. Its fediment, when placed on hot iron, turns black and fœtid.

It contains an alkaline falt, together with a calcareous earth, and an oily or bituminous matter.

Its virtues are fimilar to thofe of the *Tilbury* water, but in a lefs degree.

Tonstein, *in the Bifhoprick of Cologne, Germany.*

This is among the moft noted waters of Germany.

The water has a brifk fubacid tafte at the fountain, which is loft by expofure to the air.

It is clear and limpid when taken up from the well, but becomes turbid by ftanding; owing to the lofs of its fixed air.

It contains a chalky earth with an alkali and a little fea falt.

Its virtues are fimilar to thofe of the *Seltzer* waters, but it is more purgative.

It may alfo be ufed with advantage for common drink, either by itfelf or mixed with wine; and that either in acute or chronic difeafes, where diuretic or deobftruent remedies are required.

TOWNLEY.

See *Hanbridge.*

TRALEE, *in the county of Kerry, Ireland.*

It is a chalybeate water, of the nature of that of *Caftleconnel.*

TUN-

TUNBRIDGE. *The* WELLS *are fi-tuated about five miles from the town of Tunbridge, in Kent.*

This is at prefent one of the moft famous chalybeate waters in England, and the moft reforted to of any, tho' it does not feem to be preferable to many others in this kingdom.

It is a brifk, light water, has a ferruginous tafte, and contains alfo a little fea falt.

Expofed to the air it foon lofes its virtues; as it does alfo in a few days in bottles.

It is ufual at times to mix with the firft glafs of the water, taken in the morning, either a little common falt, or fome other purging falt, in order to make it operate by ftool. If the ftomach be foul, it is apt to vomit.

It is chiefly reforted to in June,
July,

July, and Auguſt; and is recommended in all thoſe diſorders in which the celebrated *Spa waters of Germany* are ſerviceable. It poſſeſſes the ſame general virtues as thoſe waters, but in a leſs degree.

UPMINSTER, *near Brentwood, in Eſſex.*

This is a ſtrong ſulphureous water, impregnated with a purging ſalt, and natron.

It retains its ſulphureous quality a long time.

It is purgative and diuretic; and in its virtues ſeems to reſemble the *Aſkeron* water.

VAHLS, *in France.*

The well is near Vahls, in Dauphiny.

The water has a briſk ſubacid taſte at the ſpring; which is loſt before it reaches

reaches Paris, for it then taftes faltifh.

It contains 455 grains of natron in
the gallon.

It is diuretic, and fomewhat purgative; and is fimilar in virtues to
the *Seltzer* and *Clifton* waters, though
lefs powerful.

N. B. Near to this is another fpring,
called *La Marie*, of the fame kind, but
weaker.

WALTON, *near Tewkeſbury.*

This water contains the fame ingredients as that of *Cheltenham.* The
only difference between them confifts
in the quantity of the purging falt
in the latter being fomewhat greater,
whilft the *Walton* water has rather
more hepatic air.

WARDREW, *in Northumberland.*

It is fituated between Cumberland
and

and Northumberland, on the banks of the river Arden.

It is the moſt cold ſulphureous water in the three northren counties. It contains ſea ſalt, and therefore reſembles in virtues the *Harrogate* water. The ſalt is in the proportion of about 22 grains to the gallon.

It loſes both its ſmell and virtues by carriage and keeping.

WEATHERSTACK, *in Weſtmoreland.*

This is a weak chalybeate water, but contains a large portion of ſea ſalt. In the ſummer it ſmells of ſulphur, but not in the winter.

It is purgative; and the doſe is two or three pints.

WELLENBROW, *in Northampton-ſhire.*

It is a ſlight chalybeate water, reſembling that of *Iſlington.*

WEST

§

WEST ASHTON, *in the parish of Steeple Ashton, Wiltshire.*

It is a weak chalybeate water, refembling thofe of *Iflington* and *Tunbridge.*

WESTWOOD, *near Tanderfley, in Derbyshire.*

This is a vitriolic chalybeate, fomewhat refembling the *Shadwell* water.

It is recommended externally for old fores in the legs.

N. B. The coal waters, in general, in this part of the country, are alfo vitriolic.

WEXFORD, *in Ireland.*

It is an agreeable chalybeate water, fimilar in virtue to that of *Iflington.*

WHITE-

WHITE-ACRE, *near Trales, in Lancashire.*

This is a very clear, brisk chalybeate water, resembling that of *Lancaster* in virtues, but it is said rather to bind than purge.

WIGAN, *in Lancashire.*

It is a clear chalybeate water, resembling those of *Hampstead* and *Islington.*

From the bottom rises an inflammable vapour, which takes fire at the surface on the approach of a lighted candle.

WIGGLESWORTH, *in the parish of Long Preston, in the West Riding of Yorkshire, four miles south of Settle.*

The water is very black, and has a
strong

ftrong fulphureous fmell, with a faltifh tafte.

Drunk to the quantity of three
quarts it purges, and two quarts are
faid to vomit, though it is rather un-
common that more fhould be required for the former than for the latter.

WILDUNGAN, *in the country of*
Waldeck, Germany.

This water at the fountain has a
brifk fubacid tafte, which it lofes by
expofure.

It is of the fame kind with that of
Buch, but weaker.

It is one of the mildeft acidulæ
known, and may be ufed as common
drink alone, or mixed with a fmall
portion of wine.

Though it is not efteemed ftrong
enough to remove obftinate chronic
difeafes, and clear the firft paffages,
yet

yet it is excellent for blunting and diluting acrid, fcorbutic, and gouty humours, when taken in large quantity, and for a fufficient length of time.

WIRKSWORTH, *in Derbyfhire.*

It is a weak fulphureous water, impregnated with a purging falt, and is alfo chalybeate.

It is recommended in fcrophulous, and cutaneous diforders.

WITHAM, *in Effex.*

This is a chalybeate water of confiderable ftrength, and is alfo impregnated with fea falt, but it will not bear carriage, and muft be drunk at the fountain.

It is very diuretic, and has been fuccefsfully prefcribed in hectic fevers, in weaknefs occafioned by long difeafe, in lownefs of fpirits, nervous

com-

complaints, want of appetite, indigeftion, habitual cholic, and vomiting;
in agues, in the jaundice, and begin--
ning dropfy; in the gravel, and in
afthmatic and fcorbutic diforders.

ZAHOROVICE, *in Germany*.

The fpring is near to this village,
in the diftrict of the Caftle of Suiet-
lovia, in a rocky valley, by the fide
of the river Nezdenice.

It is an acidulous water, falter, but
lefs acid than that of *Nezdenice*; and
it is alfo fomewhat pungent and fœtid.

It is in great efteem in the neighbourhood, particularly for the cure of
fcrophulous diforders.

CON-

CONCLUSION.

FOR the fake of brevity, I have omitted a particular defcription of each water in the preceding account, and occafionally referred the reader to fome water of the fame kind which has been more fully treated of; and the general virtues of the different claffes of waters are alfo defcribed at large in the Introduction.

In the Appendix to Dr. Prieftley's tract, I have given directions for imitating fome of thofe waters. The acidulous waters of the 5th clafs, for example, may be imitated, and even excelled, by fimply impregnating water with *Fixed Air*. The folid ingredients are known to be of little or no confequence. If, however, thefe be de-
fired,

fired, they may be added in the proportions directed under the article *Seltzer
water*; though it is by no means neceffary that thofe proportions fhould
be ftrictly adhered to.

A purging water, anfwering perhaps all the intentions of thofe of the
6th clafs, may be made as directed for
the *Seidfcutz water*. Rochelle falt,
or vitriolated natron, may be fubftituted for the vitriolated magnefia, if
the latter be too naufeous; and a little common falt may alfo be added.
If the water to be imitated be a falt
water, like that of the fea, the common falt fhould be in the greater proportion.

The chalybeate waters of the 1ft
clafs may be elegantly fubftituted, by
water impregnated with Fixed Air, in
which iron-filings, or wire, has been
infufed: or they may be made as directed under the articles *Spa* and *Pyrmont
water*.

water. The chalybeate purging waters of the 2d clafs may be imitated by adding to a gallon of this water two or three ounces of vitriolated magnefia, or other purging falt, and, if you will, a little fea falt.

For the fulphureous waters of the 3d clafs, water impregnated with hepatic air may be advantageoufly ufed : or they may be made as directed under the article *Aix-la-Chapelle water.* If they be alfo required to be chalybeate, or purging, or both, iron-filings, or vitriolated magnefia, or both thefe, may be added, together with a little fea falt, according to circumftances. For cold fulphureous waters, both fixed and hepatic airs are to be employed, as mentioned in the Appendix ; and even for the hot fulphureous waters it may be proper to put a fmall proportion of chalk with the fulphurated

rated kali into the lower veffel A of the apparatus.

They who have a knowledge of natural philofophy, will perceive that thefe artificial waters are not only equal, but even fuperior to the natural ones, efpecially when they cannot be drunk at the fpring head. Their vir- tues, for the moft part, depend on their *volatile* principles, and art can make water imbibe more than double the quantity of fixed, or hepatic air, that the ftrongeft natural waters are ever found to contain. The latter are alfo frequently impregnated with hurtful, or, at leaft, ufelefs ingredients; and we cannot always be fure that we have them genuine. It is not, how- ever, by any means, the Author's wifh to profcribe the ufe of the natural wa- ters. Many of them have particular virtues, as has been proved by un-
doubted

doubted experiments: and there are others which art cannot yet sufficiently imitate.

Many people again, through prejudice, will not use the artificial waters, as they do not believe it possible that they can be made sufficiently to resemble the natural ones; but even these will not object to the use of water impregnated with *fixed* or *hepatic air* in a *medicinal* view.

Water impregnated with fixed air is now known to be a very powerful antiseptic, or corrector of putrefaction. It will preserve flesh kept in it sweet, and even restore it after it becomes putrid. It is therefore given with great success in putrid fevers, in the sea scurvy, in dysenteries, in mortifications, and in other disorders arising from a putrid cause, or attended with putrefaction, a draught of it being taken now-and-then, or even by way of common drink. But the ingenious

genious Mr. Bewly has invented a ftill
better method of exhibiting fixed air,
as a medicine : he directs a fcruple of
kali to be diffolved in a fufficient quantity (fuppofe a quarter of a pint, or
lefs) of water, which is to be impregnated with as much fixed air as it can
imbibe; this is to be drunk for one
dofe *. If immediately after it a
fpoonful of lemon juice, mixed with
two or three fpoonfuls of water, and
fweetened with fugar, be drunk, the
fixed air will be extricated in the ftomach; and by this means a much
greater quantity of it may be given
than the fame quantity of water alone
can be made to imbibe. In this way
I have given it in the above diforders
with the beft effect.

But for the important difcovery of
the efficacy of this medicine in the
ftone and gravel we are indebted to

---

* Mr. Bewly directs it to be prepared in larger quantity
at a time, (as indeed it ought, in order to fave trouble)
and calls it his *Mephitic Julep.*

Mr.

Mr. Benj. Colborne. After long un-
dergoing the fevereft tortures, unre-
lieved by other remedies, of which he
tried all of any repute, he firft expe-
rienced its happy effects in his own
perfon, and afterwards recommended
it to many of his fellow-fufferers with
the fame fuccefs. For a full account
of its effects, with a variety of cafes,
fee Dr. Falconer's Appendix to Dob-
fon's Commentary on Fixed Air, from
which the following manner of preparing and ufing it is tranfcribed.

" Put two ounces and a half troy
weight, or two ounces and three quarters avoirdupois, of dry falt of tartar
into an open earthen veffel, and pour
thereon five full quarts, wine meafure,
of the fofteft water, that is clean and
limpid, that can be procured, and ftir
them well together with a clean piece
of wood. After ftanding twenty-four
hours, carefully decant, from any in-
diffoluble refiduum that may remain,

as

as much as will fill the middle part of one of the glafs machines for impreg- nating water with fixible air. The alkaline liquor is then to be expofed to a ftream of air, according to the di- rections commonly given for impreg- nating water with that fluid. When the alkaline folution has remained in this fituation till the fixible air ceafes to rife, a frefh quantity of the ferment- ing materials fhould be put into the lower part of the machine, and the fo- lution expofed to a fecond ftream of air, and this procefs repeated four times.

" When the alkaline liquor fhall have continued about forty-eight hours in this fituation, it will be fit for ufe, and fhould then be carefully drawn off into perfectly clean bottles (pints are, I think, preferable) and clofely corked up. The bottles fhould then be placed with their bottoms upwards in a cool place; and with thefe pre- cautions it will keep feveral weeks,

and

and perhaps much longer, very good.

- - - - - - - -

" About eight ounces by meafure have been taken thrice in twenty-four hours, and have agreed well with the appetite and general health; but I apprehend moft people will think this too large a quantity; and I believe, that for moft cafes, two-thirds of a pint of the alkaline liquor in twenty-four hours may fuffice. Should the bulk of the feparate dofes be thought too large, the alkaline folution may be made of double the ftrength; in which cafe half the quantity will be enough.

" The times of taking three dofes in the day, have been, I believe, pretty early in the morning, about noon, and about fix in the evening. If twice a day, about noon and in the evening; and if once, which in many cafes feems fufficient for a preventative, about an hour and a half before dinner. Common prudence dictates, that fuch a

remedy

remedy fhould be taken at fuch times as the ftomach is leaft likely to be loaded with victuals.

" I do not find, from obfervation or inquiry, that a rigid adherence to any particular regimen of diet is neceffary, farther than the ufual prudential cautions of moderation and temperance.

" The reverend Dr. Cooper has made ufe of fruit, wine, and other things fubject to acefcency, during the time of his taking the folution ; yet no perfon has received greater benefit. I however think it would be advifeable to abftain from acids, and from fuch things as are fubject to become acefcent, for fome time before, and alfo after the time of taking the dofes of the alkaline folution."

Mr. Bewly found his head affected by a dofe which he took, which alfo proved a pretty ftrong diuretic : but it was a very large dofe, containing twenty-four ounces by meafure of

fixed

fixed air. In general it has no perceivable effects. If it fhould prove cold or flatulent to the ftomach, Dr. Falconer recommends a fmall portion of fpirit to be added. He fays too, that hot milk in the proportion of about one-fourth, is a very grateful addition, efpecially in cold weather, and tends much to reconcile it to the ftomach.

When the lungs are purulent, fixed air mixed with the air drawn into the lungs, has repeatedly been found to perform a cure.

The bark may be given with advantage in water impregnated with fixed air, as they both coincide in the fame intention.

Fixed air may be applied by means of a fyringe, or otherwife, to putrid ulcers, mortified parts, ulcerated fore throats, and in fimilar cafes, and it has been found to have remarkable efficacy. It may alfo be given internally at the fame time.

In

In putrid dyfenteries, and in putrid ftools, fixed air may be given by way of clyfter, as hath been obferved by the learned and ingenious Dr. Prieftley (whom I have the honour to call my friend) in the former part of this work. Fermenting cataplafms are of fervice chiefly as they fupply fixed air to the part.

In cafes of putridity, fixed air has been fuccefsfully applied to the furface of the body, expofed to ftreams of it. And there are other cafes in which it has been found ferviceable. Water impregnated with it is alfo an excellent cooling as well as ftrengthening beverage in hot relaxing weather, and it has befides the advantage of being pleafant to the tafte.

The virtues of water, impregnated with *hepatic air* may be collected from what was faid in the Introduction, concerning fulphureous waters.

F I N I S.

www.ingramcontent.com/pod-product-compliance
Lightning Source LLC
Chambersburg PA
CBHW020501270326
41926CB00008B/693